Chronature

Chrono Spatial Dynamics
and
a Geometric Theory of Matter and Light

by

Walt Froloff with Joel Froloff

Copyright 2009, 2021 by Walt Froloff

All rights reserved. No part of this book may be reproduced or transmitted in any form or by any means, electronic or mechanical, without permission in writing from the publisher.

Published by
patentAlchemy Press,
Aptos, Ca

Cover Illustrations: Joan Rose Staffen

Dedication

I wish to dedicate this book to all those who gave me blank stares and vacant looks when I tried to explain this theory. To those who listened without understanding but whose patience and listening encouraged me to dredge this from the chaotic muck inside and into the light, respectfully submitted for your perusal. But most importantly, I would dedicate this theory and book to my son Joel, who cajoled, irritated encouraged, motivated, enabled, and edited this evolving piece of work. Without him, I would not have finished in this universe's time cycle.

What's the Matter with SpaceTime?

"Not everything that counts can be counted and not everything that can be counted counts."

                         A. Einstein

TABLE of CONTENTS

Dedication	5
What's the Matter with SpaceTime?	6
Preface	9
Introduction	21
1 – Geometries of Spacetimes	131
2 – A Least Two or More Spacetimes	145
3 – Atomic Structure	205
4 – Electrons and Electron Organization	255
5 – Nucleons and Nucleus Organization	307
6 – Spacetimes in Collision	329
7 – The Intersection of Light and Matter	347
8 –Chrono Spatial Dynamics and Metric	407
Appendix	437
So what? It's yet another theory	439
Bibliography	455

Preface

Time, the final frontier... if I find a curve in time, can I say that curve is mine? Before Einstein, scientists erroneously believed that space was flat, time was perfectly uniform, no curves, and Euclidian geometry prevailed in the universe. These ideas alone kept mankind in the dark and without understanding gravity and space for a few millenniums. From Einstein's theory of gravity, we know that matter warps spacetime, in which space and time are curved in the vicinity of matter and the degree of curvature depended on the matter's density, atomic composition, and size.

Perhaps because we are Earth-bound creatures we unfortunately still think of gravity, and hence spacetime curvature as more or less static and that somehow envelops and includes all matter. We still use Newton's formula to calculate the strength of gravity on different astronomical bodies and orbits.

In Einstein's theory of General Relativity space is as dynamic as matter[1], "it moves and morphs." Smolin writes "geometry is constantly changing because matter is constantly moving. The geometry of space is not like a flat, infinite plane. It is

---

[1]   Lee Smolin, *"The Trouble With Physics"*, pg 4, pg 41

like the surface of the ocean – incredibly dynamic, with great waves and small ripples in it." Some physicists believe that space is foam and that the dynamics occur at even much much smaller and smaller dimensions. We proffer that space in the proximity of matter has this dynamic character, and flat space not so much.

From many standards of physical understanding, the field of Physics has suffered a long stagnation period from Einstein's GR. GR introduced the equality between matter and energy. So to make up some physics slack, we will go much further and proffer that matter and spacetime are the same thing, and that energy is the transition of matter into other spacetime curvatures, that matter is just another kind of space and spacetimes and that they differ by the geometries that comprise them. We will repeat that in this book as it is the major underpinning of Chronature.

Imagine a big chunk of matter, a planetary body, call it Pluto, just hanging out around the Sun. Every day it orbits into a new position in space, the space which was flat space before Pluto arrived now becomes warped or not flat. Likewise when that Pluto leaves that position for greener spatial coordinates, the warped space where it was, becomes flat again. That tells us that spacetime is dynamic on a macro scale. Spacetime changes and is determined by matter's proximity and density at a particular point in spacetime. The curvature of spacetime was warped and then became unwarped, back to its original flat and happy state. Naturally, we wish to know how fast this transition occurs and all of that has to do with time. But more than that, we will discover that the spacetime comprising matter has been overlooked.

And this is the space and spacetime to which we will direct our focus because here matter **is** spacetime organized a little bit differently from the normally taught spacetime. As a matter of Chronature theory, matter and known forces is a type or result of spacetime curvatures. These physical spacetime curvatures are the essence of mathematical abstractions of fields and forces much like physical warped spacetime provides gravitational fields and forces which can be measured by instruments and predicted through models.

Our first guess would be that spacetime warping changes at the speed of time regardless of the degree or direction of spacetime curvature. To answer the questions is there and what is time curvature physically? We must first understand the nature of matter and the mathematics of curvature from geometry.

We know that space and spacetime can be curved because of the classic proof of starlight bending around our Sun, the experiment was a real physical reality proof that Einstein's Theory of Gravity, aka General Relativity (GR) was physically correct, and not just some mathematical physics mumbo jumbo hocus pocus. Another proof is the unexplained by Newtonian mechanics for the orbit of Mercury. Astronomers quickly found that this could be useful in predicting and understanding observed phenomena. Sadly GR has not helped much at the more local levels of materials, chemistry, solid-state, condensate state physics, and most other fields from a practical standpoint. We know that below 8-nanometers, the frontier between classical or

Newtonian physics and the more philosophical quantum world, physics models start to break down. We will need something better or closer to reality to make the next series of breakthroughs in physics and chemistry, and to "bridge" GR to QM where and if possible.

A related question in all this is, can time be bent or curved? A corollary question is can time be curved to the extent as to cause time reversal? One GR Einstein Field Equation solution says it can be but what does a bend in time even mean physically? Moreover, what can curve time and how can this be done? Furthermore, as in the gravity force, can a bend in time be a straight line simultaneously in other geometries? GR teaches that space and time can be and are bent or curved in the proximity of matter. Some geometries teach that a geodesic or curved line is a straight line. These all depend on the curvature of space and the governing rules of the geometry of that point in spacetime. But what about straight monotonic time? Can time be bent to "orthogonal" and yet straight too, curved as a line in space and can it be bent beyond 180° or some other arbitrary "breaking" or "bending" limit?

The question of whether time bend can cause time reversal or send an object "back in time" is not that abstract. Since we know that time bends in the proximity of matter, time-reversal may therefore occur near or in the matter. And indeed it does as some have found.

There has been little physical evidence of a time-reversal, which by the way would violate the 2nd Law of Thermodynamics. Nevertheless,

Feynman speculated[2] that the creation of a positron or annihilation of an electron-positron pair can be explained by time reversal. Perhaps therefore time bend and even time reversal are related since it exists in nature for at least those very short durations speculated on by Feynman on anti-matter. But if very short in duration, why should that even concern us? An answer is, because as we get to very small things, like atoms and subatomics, then very short durations become infinitely long on their scale, and amplify as they scale up. We will explain how this happens and the relevance of this in a Minkowski diagram in a later chapter.

But science doesn't sleep and it's propelled by theories, so we will indulge in a new theory, a practical theory about matter, and we will start from nothing, or only something that we know has been proven correct, not speculative or conjecture. We know spacetime can be warped and does warp in the proximity of matter. Therefore space can be warped and time can also be warped. Moreover, if time were bent a full 360°, what would spacetime look like? No additional spatial dimensions are necessary. If time somehow underwent a Mobius transformation, or a mathematical inversion, or some other mathematical hocus pocus, then frequency-time inside the boundaries of a separate alternate spacetime would result as the two-time dimensions could not co-exist without a boundary of separation. This boundary would have a dynamic 3D surface. This in all likelihood happened once upon a time, in the spacetime proximity of the Big Bang. So this brings

---

[2] QED, Richard P. Feynmann, pg 98, "Every particle in Nature has an amplitude to move backwards in time, and therefore has an anti-particle."

us to our theory that we have in "small" measures that time is circulating continuously inside every formed atom in the Universe. This would also open the door to different kinds of matter categories, for example, dark matter.

As it turns out, mathematically, bending or math speaking transforming a straight uniformly monotonically increasing time line 360° to a frequency-time, can be done by an operation called Inversion. What physically happens to a spacetime where the time dimension is inverted? We get frequency instead of straight monotonically increasing time. That is time becomes the reciprocal of time. Our inverted time clock measures cycles instead of seconds and our spacetime may physically take on a vibratory character. This is our first clue that we are now outside of the traditional Minkowski spacetime diagram. But mathematicians have long ago constructed the underlying scaffolding to sculpt out the structure of physical phenomena and so Chronature will make use of basic well-understood mathematics to construct the physical models necessary to explain the data that science provides.

Some believe the mathematicians have hijacked physics with their hard-to-understand mathematical methods which elude us because they are based on non-physical models or have a tenuous connection to physical reality. Simply naming variables assigned to properties measured can be a major misleading as to the actual physical nature that that variable. Particles and attributes are given pithy names like "strange", "up", "down", "God particle", etc. Most entities are just given the status of "particle". But if the experimentalist can create a particle through a collision, without regard to the makeup or structure of that "particle", that is generally considered evidence that the theory is sound. However, we will adhere as close as possible to the physical reality that we know without automatically labeling a measurement a "particle".

    Treating time as a straight line arrow, Arthur Eddington's and Hawking's speculations about the arrow of time led them to conclude that in so far as physics is concerned, time's uni-directional arrow is a property of entropy or randomness. The arrow of time forward only indicates the direction of progressive increase of the random element, entropy. But if time were curved, onto itself, then its forward path would be reversed to some degree, hence no increased entropy. Or vice versa, decrease entropy and reverse time. Eddington's speculations and reasoning did not go beyond the randomness which became the foundation for entropy. But what is entropy, where does entropy start, and can it be reversed? Does it start inside the atom, quark, electron? I proffer that whatever entropy is, it starts inside the atom not outside as Eddington presupposed because time inside the atom is different from time outside the atom and is order-preserving. Einsteins GR tells us at least the much, there is a proper time and then there is the observer's time, and they are not the same. The theory of Chronature will test the mechanics of time inside the atom and prove that curved onto itself becomes frequency, frequency-time creates a different kind of spacetime and that spacetime comprise the inside of an atom giving it all the properties that we see and measure. And at no time did my hands leave my wrists.

Most will agree that currently, the models for matter are inadequate, as mostly statistical and thus probabilistic methods are employed. This starts with something not entirely understood, called entropy. The macroscopic definition of entropy, an expression of the disorder or randomness of a system, for a thermodynamically reversible process as

$$\Delta S = \int \frac{dQ_{rev}}{T},$$

Entropy behaves as a function of the state in the spacetime where the measurement occurs, modeled as a consequence of the second law of thermodynamics. Absolute entropy is defined using *statistical mechanics*. The most elemental variable of this thing called entropy is Time, and is in accordance with the Eddington model, to have a forward direction or arrow of time. Because matter seems to disperse in space and time, mostly entropy is "found" in the universe to be increasing. But since we don't know what most of the Universe is made of, talking about Dark Matter and Dark Energy, we have no real clues as to what is going on there and we rely on unproven theories, conjecture, and speculation. Taking matter as composite individual spacetimes in a judicious application of a physical model, the Chronature model provides some order and structure in contra distinction to the randomness and statistical nature of current models and theories.

The Chronature model would appear to violate the 1st and 2nd Law of Thermodynamics for energy production and system entropy, but not the Prime Conservation Law of Spacetime Curvature. It will also seem irrelevant for some very important tasks such as cracking water. As a result of Temperature's irrelevance in this kind of an ordered system, the 3rd Law of Thermodynamics will appear in violation as well, where random statistical states generally rule but which can be coxed out of some randomness and leveraged to reduce bond strength for minimum effort separation. The 3rd Law is also a peculiar one, as it states that the entropy of a system approaches a constant as temperature goes to absolute zero. So a denominator, temperature, which approaches zero generally results in the entropy approaching a very large value, something like infinity. The argument is that the delta heat out also approaches zero and the integral is thus a constant. And this is a very strange formula, as we don't know what entropy is or even what absolute zero temperature is, except by convention. This just compounds our ignorance of the very basic tenents of science and shows that our current models of matter are grossly inadequate.

     One of the many great mysteries in physics is why our Universe is made of matter and why is there not enough to explain the rate of expansion. Which also begs the question of what is the nature of spacetime expansion? According to some theories, the same amount of matter and antimatter was produced during the Big Bang. When matter and antimatter met, they are annihilated, early on of course. As the Universe cooled down, it was expected that all the matter would eventually "find" the antimatter and annihilate all of both. However, and fortunately for us, there appears some matter and no antimatter left.

What we know for certain is spacetime curves around matter. Hence some physical spherical spacetime constructs can be represented by mathematical constructs for example quaternion mathematics, with its constant real value as frequency-time and the 3 complex components for spatial dimension. The constant real value in the quaternion is time and provides what is known physical property as rotation inside an atom. So Chronature will start with spacetime within the atom must have only 4 dimensions just like spacetime outside of the atom, 3 spatial and 1 temporal. Remaining true to what is known about spacetime, our model of spacetime will have only 4 dimensions. We will relax the rigidity of geometry to alternate spatial dimension geometries, flat Euclidean, non-Euclidian Elliptic, non-Euclidian Hyperbolic, Spherical Non-Euclidean, and various time dimension forms from the set including uniform, monotonic, non-monotonic, inverse, negative, frequency, etc.

There exist theories speculating higher dimensions exist above the three we know and experience. Specifically more spatial dimensions, as postulated in String Theories and reported by Randall[3], "if there are extra dimensions..". These are in direct contra-distinction to Chronature which holds that extra spatial dimensions exist but in separate spacetimes, in the classical definition of 4 dimensions per spacetime. Randall further admits "Once again we are faced with the question of how to connect a beautiful symmetric theory to the physical realities of our universe." Chronature proffers that

---

[3]   Lisa Randall, Warped Passages, pg 300, pg 352

reality is structured in spacetimes of 4 dimensions each and that higher dimensions do not exist. Furthermore, those higher dimensions cannot be proven by anybody because the hypothetical small dimensions in question require the measurements to be made at the Plank scales of 10^{-33} cm, something we will not be capable of doing for some time if ever in our relativistic time dimension.

Let's ponder for a minute on how the Chronature model of reality all happened. The models for the creation of matter, atoms, and atomic spacetimes must begin where time begins and the 2nd Law of Thermodynamics did not yet exist. Hence the bending forward arrow of time ostensibly started at the Big Bang and continued outward into the spatial dimensions. So once upon a time, time had to reach the edges of a positive and negative infinite universe, something that is still happening. On its way, time-space shear formed eddies of spacetime. In a zillion repetitions time curved completely onto itself into a sphere and formed first subatomic time vortices which then coalesced into larger time vortices of all different size individual spacetimes.

Consistent with our knowledge from Einstein's General Relativity, space as we now know obeys the rules of the geometry that it subtends in the proximity of matter. But Einstein's GR stops at the boundary of matter in general, and the atom specifically. And Chronature begins and lives there, a beginning at the end.

Introduction

There are many very fundamental unanswered or poorly answered questions and inconsistencies upon which modern Physics and Science rest. Some of these questions are:

Why does matter distort spacetime?
What is matter ?
Why can't we create matter from energy even though we know from $E=mc^2$ that they are interchangeable?
How does matter warp spacetime?
Why does matter warp spacetime?
Does an atom age and how old is it?
What is the physical reality of an electron "cloud" inside of an atom?
What's the atom's boundary made of?
What is an atom made of, ie what are electrons, protons, or neutrons aside from pieces of sub-matter for which we have names but little understanding as to their physical nature?
What is a charge?
Why are there "positive" and "negative" charges?
What is the physical structure of a photon?
What is the structure of EM or photons?
What is the physical nature of magnetism?
Why is magnetism?
Why does magnetism have 2 poles, north, and south?
Why is light's trajectory always a straight line?
Why does light travel at speed C?
Why does the angle of incident light equal the angle of reflection?
What is mass and what is its relationship with matter, yes they are not the same thing?

From a physical atomic model standpoint, what is parity or spin?
What is the physical nature of a quark strangeness, color, up/down?
Why are there 6 quarks and vary per nucleon?
What are quarks and what are they made of?
How could something as small as string theory's mathematical speculations of almost dimensionless membranes physically comprise something as large as quarks?
What is the relationship between the dimensions of space and the dimensions of time?
Why do nucleons congregate in the atom against the repelling forces of the proton's positive electrical charges, ie what is the strong force in a physical model of the atomic structure?
What is the equivalence of matter and energy in physical terms?
When is matter converted to energy, what is physically occurring? ie, what is the mass defect that physicists talk about in actual physical terms the conversion of mass defect for the kinetic energy of particles and EM?
How fast does time travel?
How do positrons and anti-particles generally travel backward in time?
How does an electron travel backward in time to look like a positron?
Why is gravity such a weak force compared to the other 3 known forces?

Chronature provides a physical model and so must explain first and predict second, a manifestation of physical reality. Probabilistic or statistical mathematical concepts are avoided as these tend to be model crutches to physical reality explanations and used primarily when a true understanding of a phenomenon does not yet exist.

For example, despite the "finding" of the Higgs Boson, our fundamentally most basic models of energy, entropy, matter, the atom, subatomic particles remain the same, wholly inadequate. The general lack of answers to the very basic questions posed above only shows that our current fundamental particle models are too primitive, immature, incomplete, or incorrect and that they should be discarded at the earliest opportunity. But even more importantly, if the answers to those questions listed above do not interest you, then you are probably wasting your valuable time reading this book. But for those armchair physicists and curious, pondering these questions are life's occupational hazard, read on.

For us here and as with General Relativity (GR), spacetime begins at geometry and specifically the mathematics of geometries. But there are many kinds of geometries, so which is the applicable geometry of space and spacetime for matter? Seems the trouble with understanding how gravity worked in the past was our over-reliance on one type of geometry, Euclidean. Thus GR in the application of space and time dramatically changed our understanding of gravity from a mysterious "action from a distance" to the geometry of space and time geometry continuum.

To understand the different geometries, we begin at the most fundamental level that a straight line is and must always be a straight line in compliance with the geometry that the straight line travels or traverses. A straight line is only as "straight" as the space that it exists in or defines it, even though it may not look "straight" to our eyes. We are predisposed, hardwired to Euclidian Geometry, and for this reason, the straight line is always assumed to be without bounds straight. This is not to be confused with optical illusions using straight lines. See below for an example of illusive straight lines that optically challenges our fundamental understanding about line perceived straightness, fooling our brain into interpreting something different from what is.

Despite the optics, the grids of lines in the figure are all straight line segments, as defined in Euclidian Geometry. If we start with and understand the straight line in the various other geometries, we will understand that all different types of space are governed by the straight line, in the situation and circumstances surrounding and supporting a straight line. Add a time with space to the continuum and the geometry becomes even more interesting.

Einstein made his monumental discovery cracking GR by discovering that world lines, lines in spacetime geometry, were straight vertical lines a distance away from matter but bent parabola looking straight lines in the vicinity of matter. When he learned about a man who had fallen 3 stories and lived, he inquired if the man felt any forces. Upon learning that no forces were felt by the falling man, Einstein concluded that straight lines in spacetime, world lines, were not like straight Euclidean lines. He is quoted as saying "this was the happiest thought" of his life[4]. The world line for inertial acceleration was curved and turned out identical to the world line for a body accelerating under gravity alone. But a world line for a body without acting forces would be a straight vertical line. Since there were no forces externally acting on a free-falling body this could only be accounted for in the Minkowski diagram if a falling object curved locus was itself a straight line. Einstein made the connection, that spacetime near matter must have a non-Euclidean geometry since the falling object world line would have a straight line

---

[4] Space and Time - Minkowski's Papers on Relativity, Minkowski Institute Press, pg 24

and any curvature in spacetime was just that, a straight line in conformance to another geometry.

Thus Einstein deduced that this "gravity" acceleration path was a straight line but was defined by another geometry, one warped with respect to the flat spacetime we always just assumed, one that was flat and uniform. Therefore the spacetime was warped or curved near matter and the proximate matter would behave accordingly in a geometrical sense. This plus knowing that acceleration is caused by one or both of two basic quantities, rate of change of speed, and or rate of change of direction. The invisible change in geometry of a moving time, the 'C' vector in the diagram, is what causes the change in time curvature, rotating an entity inertial reference frame time component thereby developing an invisible space component in that entities local spatial axis manifesting a physical acceleration without additional external force.

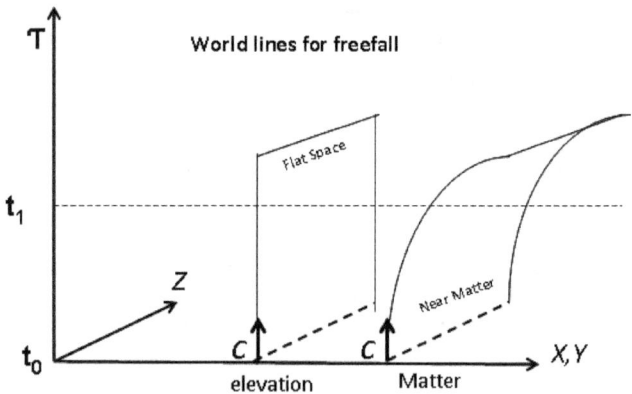

Although Einstein never knew why spacetime warps in the proximity of matter, he would forever break us out of our hardwired acceptance of flat aka Euclidean spacetime everywhere and into understanding that nature does not play favorites with geometries, and that nature chooses how and where it will define spacetime in whatever geometry it deems fit.

GR rests on the irrefutable fact that a straight line in physical yes-we-can-see it space can bend and still be a straight line. It appears very non-intuitive and inconsistent that this should be so but is nevertheless in perfect accordance with a non-Euclidean Geometries Hyperbolic-Lobachevskian and Elliptical-Riemannian. Our historically relatively new understanding of Lobachevskian and Riemannian geometries leaves us with some anxiety because everything we learned about a straight line in Euclidean Geometry screams NO!, if a line is curved it cannot be straight. Not knowing better, most of us are hardwired to believe Euclidean Geometry subtends and prevails in all spatial and temporal realms, and that non-Euclidean geometries are a mathematician's wet dream, not applicable for most situations and circumstances outside the classroom.

But in accordance with nature, this geometrical novelty determines a known physical reality in the proximity of matter as GR holds that a straight line is bent or warped in the vicinity of matter, this is still a "straight line", and matter bends or warps the spacetime proximate. Einstein did not know why matter caused spacetime warpage, only reasoned that it explained the presence and existence of a physical acceleration we call gravity.

This begs the question to what extent a straight line can be bent and remain straight? Can a line kink at a right angle? Yes in Discrete Geometries but no in analog non-Euclidean Geometries. Can a straight line bend into a smooth closed polygon? The answer is YES in Spherical, Riemannian, or Elliptical Geometry but NO in Euclidean Geometry. Since nature does not play favorites with geometry, we should expect that all analog geometries exist in nature. By relaxing the 5th postulate, Geometers have been able to discover these other mathematically consistent and true geometries. The best we can do then is to model spacetime with mathematically consistent geometries.

At the cost of being too metaphysical but so that we don't lose sight of the importance of discussing the straight line in physical space, please note that space is comprised of straight lines in three orthogonal dimensions, and time in a fourth dimension. Anything derived from understanding the straight line mathematically, including its bends and curves, therefore brings us that much closer to understanding the physical nature of space, spacetime, and finally, matter.

Some straight lines:

1   2   3

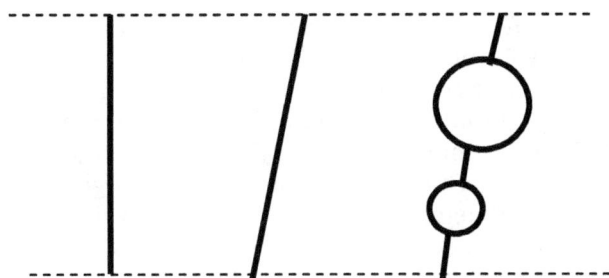

Although they may not all appear so, these are straight lines in 1) Euclidean, 2) non-Euclidean Hyperbolic Geometry and 3) non-Euclidean on the Hyperbolic and Elliptic composite geometries on the surface. Great circles or geodesics are the equivalent straight lines in Elliptic non-Euclidean space. So all of the line segments are geodesics, the straight-line distances between points are the metrics. "Straightening out" these lines from their respective geometries and comparing to them in an equivalent Euclidean geometry, we can by inspection see the reason for space distortion in the vicinity of matter that we experience. In the composite geometries of spacetime, 3, comprising Hyperbolic and Elliptical Geometry, adding the straight line segment lengths from the governing geometries, where the geodesic seen as the circles, are unbent into their flat space straight line segments, we can see that the total length of the straightened but flat space straight line segments are longer than the same straight line in flat space, Euclidean. In this way, nature draws longer spatial lines through matter, where atoms are comprised of the elliptic geometry spheres or circles in 2D and hyperbolic curved space in between. Hence straight lines, geodesics, and therefore space is longer or more expansive respectively than they should be found in flat Euclidean space away from matter. Hence the spacetime warp occurs in the proximity of matter and complies with the non-Euclidean Geometry, making the straight lines longer relative to the Euclidean Geometry of straight lines of flat space, causing all of the 4 known physical forces. This is illustrated in Figure 1. Amazingly enough, this invisible geometry change of spacetime curvature is what causes the very real physical force of what we know as gravity. It follows that since spacetime wrapping around other geometrical entities causes the very real force of gravity by just making the straight line longer relative flat space Euclidean straight lines, it must then be capable of causing other real measurable physical forces where the "wrapping" is tighter and smaller. That is to say, that spacetime curvature geometry of matter will stretch and compress an entity near or inside it like gravity does because it is just a much smaller and more varied kind of field, for which the spacetime geometry differs at every point. And this, spacetime geometry changes will appear as acceleration fields near and inside of matter, starting with an atom and giving us "electrical", "chemical", "magnetic" and "nuclear" forces that we measure. The quotations are added to show that these are just names for entities that we will define from a standpoint of spacetime geometry alone.

Not surprisingly, the "straight lines" and geodesics responsible for the curvature as we have drawn them are not visible, and the physical reality of straight lines in the geometry of spacetime is therefore also not visible. Although we know all of this, our training nevertheless still compels us to believe straight lines and space are all Euclidean and that distances are Pythagorean. However, because matter is made of nothing but curved space, of curved straight lines or geodesics, these curved straight lines are longer than those in flat space, this spacetime warping near and inside matter, bounds other kinds of spacetime metrics, or distances.

What is a spacetime warp?

Chronature begins where Einstein's GR leaves us, that spacetime is warped in the vicinity of matter and that matter warps spacetime and this results in what we perceive as the "force" of gravity. The figure below illustrated by Epstein[5] shows how warped spacetime would cause an object to "fall" toward matter. A body with a locally inertial reference frame at elevation wrt matter would through the passage of time alone, acquire a spatial acceleration component, ie gravity. To repeat from above, this happens through an entity's axis rotation of its local inertial reference frame aligning with the warped spacetime continuum. This continuum envelops all matter, creating longer spacetime lines and causing spacetime "distortion" or warp which embeds the alternate spacetimes of atoms. These are all connected in a continuum of mathematically consistent straight lines determining a 4D time and space unity. Since there is no free spacetime lunch, this is done at the expense of the local inertial frame time component, which "slows local time" and gives us fewer local time ticks.

---

[5] Lewis Carroll Epstein, *Relativity Visualized*, pg 150

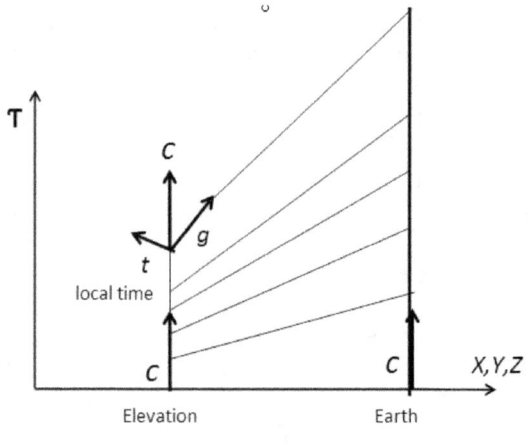

FIG. 1.1

Thus gravity, as discovered by Einstein, is an artifact or property of spacetime curvature, and that the physical nothingness of spacetime curvature alone, creates something as real as the physical force of gravity which is measurable and repeatable. One has merely to lose our footing to feel the very real physical force of gravity's acceleration. Following spacetime curvature geometry rules and reasoning will lead us logically to the result that spacetime curvature alone can account for not only gravity but the other three known forces in the Universe: EM, Weak Force, Strong Force, and yes even the Higg's Field. Therefore geometrically speaking, atoms themselves are, yet to be proved, nothing more than artifacts of spacetime curvature.

The diagram shows that the spacetime continuum has a coordinate system that is stretched from Euclidean, flat, to a stretched new coordinate system. And everywhere in between those is another slightly different geometry. The grid, represented by the orthonormal 2D crosshatch of lines of space and time, is the geodetic, straightest possible lines, between points in such a space. Mathematically that is what's called Geodesic Deviation which creates and sustains the acceleration fields responsible for the measured characteristics and attributes of the interacting entities. Spacetime geometry or warpage is again, is what causes the dynamics of motion telling entities in the spacetime how to behave.

Furthermore, although spacetime warpage is brought to us by the invisible, the abstract artifact of alternate geometries, the results from the application of the mathematics of this curvature geometry are as solid as diamond, as real and powerful as all the forces of nature. And since the four known forces provide acceleration without an external force acceleration, they are all artifacts of spacetime curvature and differ mostly by the degree and direction of curvature magnitude of spacetime.

Geodesic deviation mathematics becomes important in the EFE and is captured in the Ricci Tensor and others. Later, we will attempt to guess/find the actual geometry and curves to simplify the calculations.

What is the speed of time, or time's rate?

Generally, when we talk about a rate, it's a proportion of a quantity divided by time. For example, a distance is equal to the rate of speed multiplied by the time spent at that speed, Distance = Rate of Speed x Time. A spacetime diagram showing the set of points regarded as simultaneous by a stationary observer will plot the distance x for the time rate c multiplied by time t. So roughly speaking the time rate is the speed of light c and is the speed of time.

In our classical notion of spacetime, generally accepted x, y, z spatial dimensions, time is like a river continuously flowing, monotonically increasing at speed c through the dimension of time. And by analogy, we never step into the same river of time twice. But the time dimension inside the atom is inverted time or rotating. A solution if this notion was put forth by Gödel. Gödel's "rotating universes" allow for "time travel" to the past and solutions are known as the Gödel metric, an exact solution of the Einstein Field Equation.

What is mass?

Below is the Schrödinger Equation[6] extended to include Relativity.

$$-\frac{1}{c^2}\frac{\partial^2}{\partial t^2}\psi = -\nabla^2\psi + \frac{m^2 c^2}{\hbar^2}\psi \quad \text{(I-1)}$$

By inspection, we can see that equation **I-1** is a type of PDE wave equation with a friction term. We

---

[6] Wikipedia, http://en.wikipedia.org/wiki/Schr%C3%B6dinger_equation

have come to call this friction term "mass". Mass is a fudging parameter of what we call matter, the quantity which interacts with spacetime as resistance or inertia, and for that reason is not well defined or well understood. Some seek it in the "God particle", Higgs Boson. But this equation is our first clue as to what mass is physically or should represent in reality, spacetime friction or inertia, resistance to motion in curved spacetime or fields.

Physicists have been searching for evidence of the Higgs boson particle, a particle that is supposed to endow other particles with mass. This search has been concluded as news from CERN in its actual measurement have confirmed such a predicted subatomic particle does indeed exist and is single-handedly responsible for all the mass that we know exists.

In the Chronature model, we may define mass as a "friction" or "viscosity" introduced by interacting spacetime curvature boundaries. Since these spacetime boundaries contain and separate what is currently called particles, their boundaries are spacetime curvatures and these change the proximate spacetime character surrounding them as they move in the spacetime pulling and tugging on each other as they move in the proximity of each other boundaries. Peter Higgs, for whom the boson was named, believes that a sort of cosmic molasses pervading space is what gives particles their mass. Predicting 3Spheres of 4D spacetime are particles with space curvature boundaries explains this "molasses" effect.

What is the benefit of a physical atom Chronature model if the QM model does indeed predict a physical reality of electron orbital's even though it does not model physical reality? Since we know orbiting electrons are physically not real things we cannot reconcile that against orbiting particles as defined by the current models of the Bohr-QM probabilistic model.

And this should worry us, because much of our current research and models rest on these very amorphous foundations, resting on primitive or dubious physical models, on-premises over a hundred years; models of the atom that make little or no physical sense; models of the atom that were adjusted by divine decree to fit data so that QM could predict electron orbital's for a hydrogen atom; so that QM could be acceptable as a theory, instead of the alternative, that Schrodinger's probability model accidentally generates the same equation as the physical model of a vibrating sphere, after some "tweaking". Using a vibrating sphere for an atom model gives us solutions that serendipitously exactly match the orbital structures and order of orbiting electrons to the harmonics of a vibrating sphere.

What is an electron and where do they live?

The current models hold that an electron lives in a probability "cloud" and the electron is a physical particle that orbits a physical nucleus. But what is that "cloud", since there is no water vapor or gas at that physical level? QM teaches that we cannot know the exact position until we make a measurement and only then do our observations somehow "collapse" the electron's state so that we can make an observation. This current model of the physical atom is perplexing because although we know it is wrong physically, it seems to accurately predict some things, hence "correct" to a degree.

Picture the speedy electron orbiting the nucleus like the Flash character, so fast that it somehow miraculously creates a 3D boundary blurry cloud surrounding or enclosing a nucleus. Calculations using a circular orbit electron hydrogen model give that the electron speed orbiting a Hydrogen atom is approximately $1/100^{th}$ of the speed of light, C. Is the atom's boundary a group of connected string loops or is it comprise of the flapping membranes of String Theory that hang around a nucleus creating a peanut shape? Can an atom be a closed system subtending many even tinier loop systems? One thing is certain since an atom exists and can be isolated, it must have a physical structure in space and time. Whatever representations from the Standard Model and whatever probability is applied from QM, do not describe or address the physical nature or structure of the electron or atom found in physical reality. The QMers have not gotten off their collective keisters to produce a model for the physical world, going on now for a hundred years. Thus QM is mathematical speculation based on a bad model finagled to equal the measured electron shell solutions which are identical to a vibrating sphere. The commonality between the Bohr model and Chronature being the physically spherical configuration of the Hydrogen atom's components. QM gets there from the Schrodinger Hydrogen Model and some "quantized angular momentum" from Bohr. How did this fortuitous accident give the correct result, electron shell structure? To explain the Bohr model's correct result QM took the wrong turn to Not-even-wrong street. QM went from perhaps plausible to weird and weirder as proponents took it upon themselves to explain and expound on the theory, which had to be decoupled from any physical reality to maintain a straight face in the explanation. Yes, the Schrodinger model appears to provide a valid orbital electron structure, but not until it was modified beyond recognition by Bohr's decree to have discrete electron orbitals in violation of physical reality. Not dealing with the physical world anymore was the deviation QM made from the realm of scientific reasoning and into pure probabilistic. Einstein's thoughts were QM carries not only incompleteness but insufficient reasonableness as well.

According to the QM modified model, Menzel[7] compares and makes note of the Bohr model solutions to a vibrating sphere that:

> "as originally discussed by Bohr and Sommerfeld, **arbitrary** rules were employed to select, from a manifold of mechanically possible electron orbits, a few stationary states of special significance. ... the model was incomplete... the mechanical problem of selecting certain stable modes from all possible **modes of the vibrating sphere** has many features in **common with quantization**. The parameters m, l and n, are analogous to quantum numbers. The distinguishing difference is that in the mechanical problems the quantum numbers appeared naturally."

Menzel goes on to write:

> "The letters m, l, and n as introduced in the problem of the vibrating sphere have been adopted because of their relation to the atomic quantum

---

[7] Donald H. Menzel, *Mathematical Physics*, pg 213

numbers. Indeed, the functions of the Phi and Theta for the hydrogen <u>atom</u> ***are identical*** with those for the <u>vibrating sphere</u>."

So what is wrong with a primitive or early model you might ask? After all, that's the old "scientific" hunch method we all progress. As Einstein said we "stand on the shoulders of giants." The answer is, a lot, because the divining and arbitrary promulgation by Bohr to force-fit a marginal physical model to the known reality, leads to the confirmation that electrons are best modeled as some kind of nucleus orbiting point particles. This lead to 100+ years of an academically coerced inferior model for which more complex theories needed to conform to a lesser foundation. Building on a bad model means much more experimental work to crack the "structure of the atom" with larger and larger atom smashers, the latest and greatest at CERN. Insult upon injury, the physically primitive Bohr model with "help" from "cloud" probability QM is used throughout our other sciences most directly Chemistry, to create theories of observations and measurements of molecular interaction in physical reality, starting from reactants and ending with products. Theories explaining the thousands of specific element properties and relationships using the Bohr model are hardly believable, and Chemists have been forced to be excessively creative by constructing very tenuous and complex models to explain physical observations and experiment results using a poor model for a foundation.

    This is reminiscent of the ancient Ptolemy model of planetary orbits made to fit the divine postulated assumption that the universe revolved around the earth, very creative but of little value to progressing science or planetary orbital mechanics. Ptolemy's work required immense complexity to explain the physical planetary motions, comets, and star orbits based on a geocentric universe. The Ptolomy model was predictive and visibly verifiable but overly complex because of the political-religious forced model foundation. Better understanding and science prevailed eventually with some bloodletting first needing to overcome the religious and political authorities of the day. That is by analogy where we are in physics, chemistry, and the basic sciences today, and for a similar reason, too primitive a model at the most fundamental level, and promulgated by authority.

Another mystery is embodied in Bode's law, an empirical rule giving the approximate distances of planets from the Sun in a harmonic ring progression. It was once suspected to have some significance regarding the formation of the solar system, but today Bode's law is generally regarded as a numerological curiosity with no known justification. But from the GR standpoint of nature's use of spacetime curvature in the generation of harmonic spacetime curvature for matter, it stands to reason that the creation of the Solar System, analogous to the dunking of a floating object into water creates ripples that travel radially outward. By analogy, the planets are much smaller objects that get caught in the wave troughs at creation, the spacetime curvature between the planets formed by Sun spacetime outward generating harmonics in the form of rings at system formation, keeping the planets geodesic orbits about the Sun. Because we may still be suffering from Ptolomy syndrome, I find it impossible to believe that Bode's Law is just a "numerical curiosity" and that a much simpler better model yields simpler physically observable explanations

Moving back to the nano dimensions, it defies physical reality that the electrons in the "electron probability density cloud" would not collide with electrons in another electron cloud, and being repulsed or kicked out of orbit would instantly bring about matter disintegration and a virtual instantaneous big de-bang. That would be the physical ramification of such a theory and yet the Bohr model persists, presumably because there is nothing better. A flaky "quantized angular momentum" modification doesn't make the Bohr model any better and perhaps makes it worse because it takes the path away from physical, reasoning, and logic to the non-physical non-logical unintuitive path of untenable models.

If not an orbiting particle, what is an electron?

Respectfully submitted for your judicious perusal, Chronature invokes a completely physical model of a physical atom, wherein electrons are simply harmonics on a vibrating spherical spacetime membrane between two spacetimes, the atomic boundary. The "added" dimensions of string theories leading up to the Theory of Everything TOE are instead to be dimensions found inside of the atom. That is to say, matter is spacetime with alternate dimensions, and no tale applied nor decree made, to explain why 5, 6, 7, 8, 9, 10, or 11 dimensions exist all "curled up" comprising the only spacetime. So spoiler alert, missing dimensions do exist, but in disparate spacetimes, embedded in other spacetimes as submanifolds.

    Since we know from GR that spacetime warpage produces gravity, it stands to reason that a magnitude of 10^{40} times greater warpage than the earth's spacetime curvature could be responsible for all of the unique atomic chemical and electrical properties of the same or similar scaling order. Alternatively, finding geometric entities that twist and turn spacetime to conform with the physical phenomena that we measure is a valid scientific approach to a better model, especially if all phenomena can be explained through one elegant model. Building our model on existing equations from the vibrating sphere mechanics and topological rules, strangely yields the same quantum numbers derived from vibrating sphere solution parameters m, l and n. But as quoted above and yet to be shown, a vibrating sphere requires no constant angular momentum assertions, assumptions, or presumptions to provide a model which produces the same electron structure prediction. The mathematical solutions of the vibrating sphere model share the same physical measurements without the necessity of any divine decrees of constant angular momentum or orbits of indefinable point particles called electrons. The current probability electron orbit model can be interpreted and used much more powerfully if we just use the geometry of the electron "orbits" as the geometry of spacetime curvature. See the figure below for the harmonic boundary of the atom as it vibrates. The electrons are simple harmonics of a vibrating sphere, whose modeshapes are shown below for the real and imaginary harmonics, and with the conventional quantum numbers and shells.

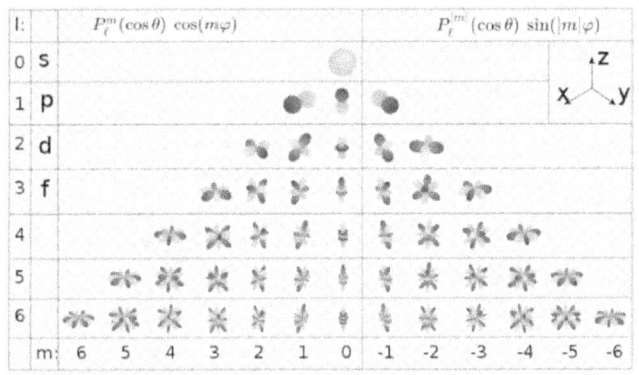

Origins and the Minkowski Space

Minkowski introduced his method of separating time from space graphically to Einstein. Typically time was the vertical axis and all of the spatial dimensions were clumped together. Show in the figure below, GR added the world light cone in the Minkowski diagram to include the variation of light speed to have spacetime curvature with variation for light speed. These causing warp and warts in the Minkowski world light cone from "clumps" and "filaments" of galaxies formed. Sparse galactic matter formations are not uniformly distributed in the Universe and are currently explained by the "inflation" theory propounded by cosmologists. We can only "see" as far into the Universe as there has been time, approximately 18 Billion years. Outside of 18 Billion light years we have no data and speculation abounds. We will not dwell on cosmological explanations although the Chronature model can be used to make much more palatable, physical analytics about the formation and expansion of the cosmos.

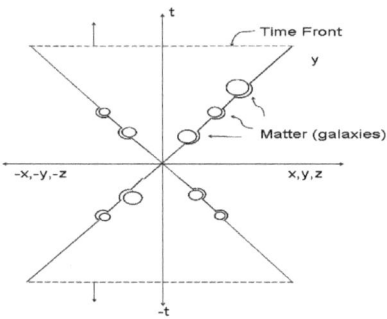

We know that the rate of time progression in our macro spacetime is C, time's relationship to space. It is not a coincidence that this is also the rate at which EM travels through general spacetime, the relationship between space and time is the rate C at which time moves. From a transition of a timeline from linear to circular, we apply that the rate of time inside an atom is C. We stipulate that to map our atomic model into a different spacetime, one in which the Minkowski time-space light cone rotates, putting us in compliance with the Gödel metric submanifold with looping time.

Suffice it to say, that the Big Bang imposed some requirements that would dictate that the curvature of time for a 3-sphere spacetime vortex. We profer that the formation of time vortexes created spacetime curvature in the Minkowski spacetime comprised other spacetimes and formed matter.

Accepting that the Minkowski diagram is substantially a correct diagram representing macro spacetime, at least as it applies to general or macro spacetime where matter resides. Embedded in macro spacetime, exist regions of non-Euclidian hyperbolic geometry with time monotonically increasing when in the proximity of matter. There is also mostly Euclidean Geometry or flat space, micro-gravity, a distance away from matters spacetime curvature effects near matter.

Macro time, embedding atomic and subatomic size 3-sphere time vortexes, the monotonically increasing macro spacetime has some time bend as well but generally unperceived by us except through an expanding universe. We speculate that this "expansion" in the Universe is obtained principally through the changes in the geometry of spacetime at the edges or boundaries of macro spacetime.

Blackhole speculation adds much to an understanding of bending time. Minkowski Spacetime can be represented as a graph of all possible trajectories for all objects in the universe. Some hypotheses predict the universe is the event horizon of a growing higher-dimensional black hole. Nothing can travel faster than light and black holes are so strong they even pull in light. So once you cross a black hole's event horizon, all future spacetime plots can only include trajectories moving towards the singularity. The spacetime graph is now just a flat line as you have no future paths. As we will see later, Gödel proved that time can wrap onto itself in Continuous Time Curves or loops. This is what is most likely going on in a black hole. Instead of a singularity in spacetime, we have a 3sphere, where time is circular and continuous.

The classic Minkowski diagram is shown below adjacent to our speculation on the nature of the time dimension in macro spacetime, adding a slight and undetectable curvature to the time dimension axis.

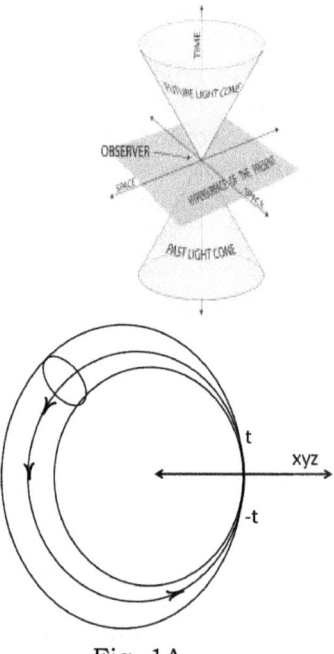

Fig. 1A   Fig. 1B

There are many models of the Universe, visible and the invisible and for different galaxies with different velocities, "redshifts/blueshifts", that are moving away/toward us respectively. Without getting into the age and cosmological nature of the universe debate the above time dimension curvature would support an oscillating universe with a small addition, wherein a hypothetical universe expansion front forms our Universe boundary, and there lies the expanding region. A space void on the other side of the front, Cosmologic Horizon, is contracting and where time is negative. The classical Minkowski diagram Fig 1A has the time axis as a straight and linear vertical axis. Our curved time axis spacetime modified Minkowski diagram Fig. 1B shows that the time axis is infinitesimally skewed but not yet measurable curved time axis, curved in a cycle on a macro scale. This begins to look like a modified Minkowski spacetime has a place in Gödel's rotating universe. The indrawn ellipse represents a world cone bounding membrane inside the world envelope which has been expanding from zero time since the big bang time front expanding universe, behind which is our formed universe and in front, the void universe or void spacetime having reverse negative time to the time zero, big bang. This defines "reverse negative time" from a physical sense. If time is thus curved in the macro universe, then the expanding universe has positive time and a receding universe has negative time. This is different from the notion of going "back in time" in the classical time sense of moving backward on the time axis towards zero and then negative.

    This idea of perceiving time as bent or cyclical answers other questions about the role of and dimensionality time. Is it real, is it fundamental? Our model has a beginning of time, the origin on the axis, where time is zero. But this is not to say that time did not exist before that, only that it has a zero value. In fact, in our fractal model, time is oscillatory and cycling forever, thus it is a fundamental dimension no question about it. But it also has a zero, where there is no time value, so it is timeless at that point. Thus Chronature is in full compliance with Relativity on this count, that the universe is finite, but without boundaries, that the universe is both timeless and timeful, and that the universe both expands and contracts.

Also from this model, the universe is static in the sense that spacetime can have an "outside" and an inside through a boundary or moving front constantly changing in its fundamental dimensions of both space and time.

This GR modified Minkowski model also begs the question of when does the time axis turn negative and can that be calculated showing that the time axis is indeed bent and the degree of the bend? Can we measure the circumference of the time bend, as Eratosthenes of ancient Greece was able to measure the circumference of the earth with a few basic measurements and an effective model? I proffer that this is possible knowing certain strategic celestial body locations and their predictable periodicities with "temporal" angles wrt the earth.

Since the spatial volume of spacetime is expanding. the positional front where the Universe is creating new galaxies and matter is moving outward, does time expand also in the same way? One can argue that perhaps it is the expansion in the time that is causing the warp or bend in the macro time dimension. And that eventually, this bend in macro time is the cause of an oscillating universe that goes through zero time ever once in a while, resetting time to zero to a "big bang" initial condition. This is now getting into the philosophical realm from which there is no return or benefit. So we leave that with the cosmologists, they seem to have prodigious quantities of time to debate those grand issues and we will concentrate on the practical aspects of understanding our physical reality.

The universe, by nature of simply existing, is "moving" at the speed of time, c. Yes, that does include you and me. We proffer that our current understanding of the universe is just too primitive or crude which provides erroneous models of space, time, and spacetime. Space and time are not separable and orthogonal and must be treated as such in our models. There are geometries, there are types and there are combinations of space and time dimensions comprising different spacetimes. The universe is comprised of a composite spacetime.

The infinites of space and time are handled by the Penrose-Carter diagram in an infinite Minkowski universe as shown in the following illustration:

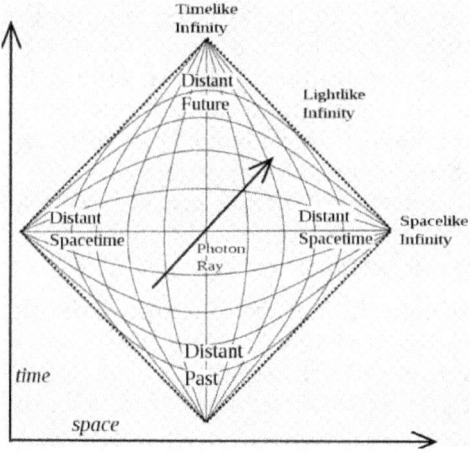

The reader will note, that the warped grid telling us that the answer is in the geometry is for the big stuff, not the small stuff. We proffer that it's the same.

    We generally know motion as a distance traveled over some amount of time. However, if distances and intervals of time are the same things, spacetime becomes more understandably a 4D construct, and that time variable is still somewhat special and apart from the spatial variables, albeit remaining independent of each other.
    Since everything is relative, things in the universe travel through a spacetime at some velocity, speed, and direction, for most things the fastest rate we call "c".

    So objects in spacetime do have different speeds and somewhat observer-dependent. South and West, for example, are orthogonal: one can travel directly to the south pole, but it's not going to affect where one is in terms of east/west, but it will affect or be affected by what time an observer will observe our moving. This affects something Einstein labeled "simultaneity".

But unlike our travel south independent of our travel west, our travel through time is affected by our travel speed in any spatial dimension by something called Lorentz time dilation-contraction. Moreover even sitting perfectly still, not traveling through space at all, our time of travel in spacetime speed is then c, "speed of light. However photons, "massless particles", on the other hand, travel through space at velocity c, they don't have a time component at all since all of their time has been given up to their space component of spacetime. Moreover, photons can never be stationary from *anybody's* perspective, as we will see later that they are spacetime transitions and special from that standpoint. Since, like everything else, photons travel at c through spacetime, that means all of its "spacetime speed" *must* be through space, and none of it is through time. So the real limitation in spacetime is time, nothing travels faster than time c.

In short, what does the Chronature model say about the dimension and nature of time?

Chronature holds that time has at least four basic time types; straight monotonically increasing time, negative or monotonically decreasing time, inverse or frequency-time, negative inverse or inverse frequency-time. More importantly, although the time dimension has a moving property it also has a geometric character. Time can bend or warp compliant to the rules of the geometry of the spacetime in which it exists or defines. So in non-Euclidean time, time could bend or warp and still be in compliance with "straight line" time. In Euclidian time, time is straight and uniform as per the Euclidian geometry. But it acts like the dimensions of space in this respect. Mostly because signs are arbitrary and reciprocals do not change the basic dimension. eg seconds and per second are different measurements of the same dimension of time. It's only how we measure the dimension of time that counts. Space dimensions are similar in that for example length units and per length units are both dimensions of space, they differ on the measurement units involved.

Why does spacetime warp in the vicinity of matter, and matter warp spacetime?

Also known as the Geometric Theory of Gravitation, GR holds that matter warps space and spacetime in the vicinity of matter. Spacetime distortion, dilation, or warp is what General Relativity teaches is the cause of gravity. As Einstein put it "matter tells space how to bend; space tells matter how to move."

But why that is so, is not known but explained by the Chronature model. What is known is that Euclidean space somehow changes to hyperbolic non-Euclidean space in the proximity of matter.

And so we lay down a postulate here, that there is only one thing that can warp space or spacetime, and that is other spacetimes. It then follows that matter is a spacetime, but a different kind or variety of spacetime. Moreover, it is the nature of the spacetime curvature comprising matter which gives all of the properties that matter exhibits, in all its different forms. Graphically we show this below for the 3 types of straight lines shown above and with the Euclidean equivalents "straightened out" by segments shown just to the right of each.

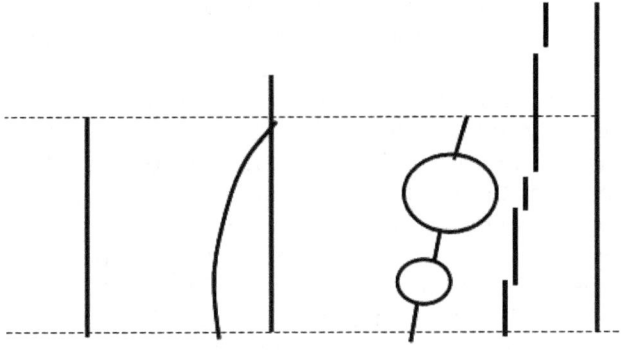

Euclidean   Hyperbolic   Hyperbolic+Spherical

We proffer that matter, like gravity, is a characteristic manifestation of warped spacetime. Matter is a consequence of spacetime in compliance with non-Euclidean Elliptical geometry. Thus the universe comprises a composite space and spacetimes. In our composite spacetime, matter is a type of spacetime having a positive curvature space (PCS) with frequency-time (FT) where time is circular inside and space conforms to the rules of non-Euclidean Elliptical Geometry. Matter, atoms, and molecules, are spheres of Elliptical geometry conforming space and spacetime different from the known non-Euclidian space adjacent to Minkowski spacetime, compliant with GR and EFE. All matter resides in the remainder of space and spacetime which is immersed or surrounded by the hyperbolic geometry curvature space monotonically progressing time, **H**yperbolic **G**eometry **S**pace and **M**onotonic **T**ime (HGSMT), spacetime, which comply with Hyperbolic Geometry for which GR applies. In a composite spacetime, space and spacetime must weave around the spherical-shaped pockets of a different spacetime or atoms, 3-sphere spacetime vortices of the **E**lliptical **G**eometry **S**pace **Frequency-time** (EGSFT), matter as we know it. This weaving in and around matter spacetime, EGSFT, HGSMT spacetime makes HGSMT spatial and temporal dimensions dilate, since the lines of space and time circumscribing EGSFT, matter, spacetime giving us space and spacetime "warp". That is hyperbolic geometry space and spacetime is more expansive or warped when it contains matter since HGSMT must circumscribe the matter radius EGSFT spacetime boundary of each atom, to occupy all spacetime other than that occupied by those atoms, EGSFT.

    In short, the excepted standard model of the atom is wholly incorrect because it rests on a model of particles orbiting particles, not geometries and dimensions which are the foundations of physical reality. The physical reality of matter exists as an atom comprising a 4D manifold within a 3-sphere of EGSFT. Where there is no matter, flat spacetime defines and can be modeled by Euclidean geometry. Whereas in the vicinity of matter, hyperbolic geometry non-Euclidean geometry space models the curved spacetime, Elliptical non-Euclidean geometry governs the spatial dimension geometric character inside atoms. These are the three composite elements that define all space, time, and matter. The speed of time inside the Elliptical non-Euclidean geometry or 3-sphere, an atom, is the same as the speed of light, C; only it exists as a circular curved or frequency form of time. So to repeat, the big difference in the two spacetime times, HGSMT and EGSFT, inside the atom, EGSFT 3-sphere, time is a frequency or periodic at rate C relative to the 3 non-Euclidean curved space dimensions inside, and monotonically increasing at rate C relative to the spatial dimensions outside the 3-sphere atom. It is this difference in the time dimension, which stabilizes the curvature of the atom in a 3-sphere and 3D elliptic spatial volume. It is the curvature of this space boundary between the HGSMT and EGSFT that dictates and defines many of the chemical properties of matter that we measure.

The atomic surface boundaries are closed diffeomorphic topological discrete spacetimes with curvature dynamics time formed and divide into two separate and distinct 4-manifolds. It is also this spacetime curvature boundary of every atom, which has the monotonic time on the outside and frequency-time on the inside, to become dilated, as well as for space to warp, as the warp is the manifestation of the lines of space which must curve around the atomic manifold spacetime 3-spheres, and hence lines of space become essentially longer and thus warped.

In the following figures, the "straight" lines are all of the lines going around the circle(s), the lines on the figure boundaries, and the circle within. Unlike the straight-line checkerboard picture above, this figure is not an optical illusion, but straight lines or geodesics in their geometries.

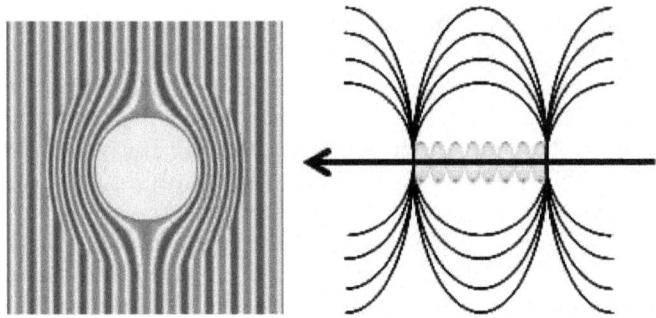

In the second figure above, the straight lines are bent around the Lorentz "squished" sphere train to show that the space dilation occurring at electrons traveling at light speed would distort the spacetime in their proximity to take on the two-pole, dipole, spacetime geometry induces straight lines that we all know as a magnet field. There is one more little thing about magnetism not explicitly shown by the spacetime lines, and that is the direction of force flux "flows" from "north" to "south". Moreover, one pole is repulsive, commonly referred to and North, and the other pole is attractive, commonly called the south by convention only.

But why would space and time transform into 3-spheres to form matter?

This is the cosmological question of the universe, regarding the formation and creation of matter. When did it happen, how did it form, what is the process and why did it happen? What happened to all the anti-matter? How come we cannot detect Dark Matter? The short answer is nobody knows, but there are theories and I offer at least one. Although any defining event can be used most theories of the formation of the universe, and matter is a big piece of that, "beginning" at some initializing event called for lack of a better term the Big Bang, Minkowski diagram Fig 1B, we speculate that at the Big Bang event, the time dimension zero event, there were only four orthogonal dimensions formed, three spatial and one temporal. The three macro spatial dimensions may have existed previously or been formed previously comprising the great void, for the release of time. The temporal dimension is the odd man out and has a rate relative to the other three, the spatial dimensions. That is although space and time are independent and orthogonal dimensions; C is the rate relating the dimensions of space and dimension of time. Hence the Minkowski Diagram graphically keeps these separate on the orthogonal axis.

Moreover, time travels at rate C with respect to, wrt, space. ie, in classical or traditional spacetime we must multiply the time displacement by the rate C, to get the equivalent distance traveled, spatial displacement, or divide traversed space by C to obtain time. Units of distance divided by units of time will yield rate, in this case, C, because time relative to classical, or macro space progresses at light speed C.

It is our cosmological speculation on the Big Bang so that spacetime curvature would be conserved, more later on the primary conservation law - Conservation of Spacetime curvature, all four dimensions required reaching infinity in the fastest route conserving mode. But the spatial dimensions, taking one-dimensional spaceline, travel essentially instantly wrt time, and hence already existed as a void. Time is the moving dimension, thus the time dimension acts differently from the time dimension in a geometric sense. ie. it doesn't follow the Pythagorean theorem. Two spatial dimensions combine as the distance between two spatial points is $d^2 = x^2 + y^2$. However, the 4th dimension "length" between two events at different places and different times are $d^2 = x^2 + y^2 - c^2 t^2$, where c is the speed of light rate multiplier. The minus sign is a big difference and clues us that we are not in Euclidean but non-Euclidean hyperbolic geometry space.

Since time has a rate of travel C wrt spatial dimensions, its introduction, at the Bang, produced spacetime "viscosity" in reaching out to infinity, time eddies or vortices formed in the space manifold, creating subatomic particles, atoms, negative frequency-time, which coalesced into larger and slightly different spacetime 3-spheres, positive and negative frequency-time, forming matter by the gravitational forces formed subsequently by their newly formed spacetime curvatures. Very tiny, subatomic, at first and then larger scaled atoms coalesced to form larger atoms than molecules, clumps, and massive heavenly bodies.

Imagine a stream of time like water moving at warp speed, at the edges where the water borders the shore, the three spatial dimensions. Time eddies or vortices are formed in the water, time, where there is shear, spacetime shear. Hence time formed eddies in spacetime to form first the tiniest vortexes of spinning, positive and negative frequency-time containers of atomic and subatomic matter were formed as time vortices within an associated 3 space dimensions to maintain a 4D manifold of a different sort adhering to non-Euclidean spherical geometry. To reach infinite simultaneously, time had to curl or bend upon itself, inverting in a Mobius or inversion transformation preserving the relativistic Lorentz translation rules, and in so doing forming 3-sphere vortexes of spacetime, spatial vortexes aka nucleons and electrons forming atoms and subatomic 3-sphere other kinds of spacetimes.

"Subatomic" 3-spheres and "atomic" 3-spheres coalesced into clumps called "matter" due to their curved surfaces we know as attractive gravitational forces, formed from the newly created spacetime curvature at the boundary of the 3-spheres. These entities coalesced under attractive or gravity forces into larger and yet larger masses followed by astronomical bodies which then formed astronomical systems and so forth. This all predates the spinning off of galaxies, nebula, and solar systems.

This simple creation order provides a simple and easy explanation of the formation of matter at the formation of the universe from the big bang event. Occam's Razor is a strong proponent of this formation theory because it explains the formation of the matter in the universe in the simplest fashion. Astronomical bodies then formed due to spacetime curvature attraction, from GR identified and accepted force of gravity resulting from the formation of spacetime warping. In this manner, matter is a mechanism for time to reach infinity instantly, in the process of forming complex 3-spheres where time inside is frequency, circling inside the space sphere volume, giving the spherical volume different properties from the other space, from which it came from the combination of only space and time. You might say that at the big bang, time took a sharp orthogonal twist in time and three turns in space to reach infinity, forming a gazillion 3-sphere like 4D Klein bottles, in compliance with Mobius transformations, one for each spatial dimension.

In other terms, the Euclidian parallel postulate was altered inside the atomic volume, to comply with elliptic non-Euclidean or Riemannian manifold geometry during the big bang universe formation conditions of spacetime expansion. In doing so, the remaining space surrounding the 4D spherical space vortexes needed to go around and had a longer path, departing from flat space and forming Lobachevskian hyperbolic non-Euclidean geometry space, leaving massive bodies of collections of these 4D sphere space manifolds, we call aggregated matter, spinning atoms coalescing into the spinning matter and then spiraling astronomical bodies.

How can we be sure that there are different types of time?

One observation we can make wrt to the monotonic or classical time that we know is that it can also be cyclical without us sensing this, as proffered above in the modified Minkowski diagram. If the positive time axis on the Minkowski diagram can bend ever so slightly, then we know that time can warp if we plot the time from the Big Bang event, the time light cone base would reach the start of time, when $t = 0$, as we can see it through our telescopes 13.7 Billion light-years away, all the way to the time front. If we think of time on the other side of the time front, 13.7 billion light-years away, the time in the void, as negative time and we connect that time axis to the Minkowski negative time axis, then time becomes cyclical in the monotonic time macro universe scale. The time front is creating matter as it travels in the positive time axis but on the other side of the time front, the universe is collapsing. Hence negative time is just positive time in a collapsing universe and not reversed time in the present expanding universe.

Thus even macro time, that time outside of all atoms, is also just cyclical time in the classical time dimension, but so infinitesimally small in curvature, that we perceive it as flat or monotonically increasing time. In a fractal universe, this kind of time on a smaller scale would be cyclical.

Suffice it to say, frequency-time has a mathematical physics foundation propounded by Gödel and mentioned in more detail below. Gödel discovered a frequency-time solution to the Einstein Field Equations which began the conversation "rotating universe", the possibility of time travel, and the existence of wormholes. But Gödel and Einstein were thinking about the large scale universe, macro time, and not the atomic scale, for their applications of rotating time solutions. Here we will adopt the rotating time into the Lorentzian Metric tensor with rotating time and non-Euclidean Elliptical space.

GR gave us system time which to a large extent implies that of succession and it appears to deprive the lapse of time of its objective meaning. However, on the existence of matter, Gödel states, "distinguishes the observers who follow ... the mean motion of matter." In his solution,[8] there is no such

---

[8] K. Gödel, An example of a new type of cosmological solutions of Einstein's field equations of gravitation, Rev. Mod. Phys. 21 (1949) 447-450

one-world time. Moreover, by making a round trip on a rocket ship in a sufficiently wide curve, it is possible in these worlds to travel into any region of the past, present, and future, exactly as it is possible in other worlds to travel to distant parts of space. "

In his published remarks, Einstein acknowledged the possibility of closed time-like lines "disturbed me at the time of the building up of the general theory of relativity." Gödel's solutions are known as the Gödel metric, which is an exact solution of the Einstein Field Equation.

Wouldn't time curvature require too much "energy" to create these very tight 3-spheres of matter altering space curvatures, precluding their existence?

Up until now, we have not mentioned energy because the word energy represents many things in many models, none of which have meaning for us at an initial mathematical level of space and time and curvature of such a formation. As spacetime curvature will be the underlying component for energy and matter we will find precise physical and mathematical meanings of the word "energy". Perhaps in today's universe, this is true, that great amounts of energy are required to form matter from flat spacetime. But at the big bang event, conditions and rules of space and time were vastly different. Energy had not yet been defined in any sense since matter did not yet exist. And that will lead us to a theory of the substance of energy and matter, and why they are equivalent to the stable and dynamic forms of the boundaries between the two different non-Euclidean spacetimes.

What are the physical mechanics for modeling spacetime curvature in a 4D?

While there is much speculation in Physics as to how nature "works", most "modern" theories have had to function outside the physical realm and therefore we will attempt to describe a physical model of an atom starting with the boundary of the atom. We know that the vibrating sphere model has the same solutions as the electron shell structures of the currently accepted Bohr-Quantum model. With the exception that the solutions to the model equations are the 3D closed modeshape surfaces, not the volume that these surfaces enclose as in the QM "probability cloud" where an electron can be found.

Firstly, the Klein bottle is a 4D creature and is depicted in the three dimensions by cheating at the handle junction. The 4th dimension, time, provides a bridge whereby the surface does not intersect but passes through the bottle via the 4th dimension. The Klein bottle is a unique mathematical entity that provides some clues as to how nature physically might be modeled using Elliptical geometry, a one-sided surface geometry but still a closed manifold. But unlike a 2-sphere, it does not have the mathematical property called "compact" manifold, a loop on the one-sided surface that can be reduced by one dimension. This has to do with the topological property called orientability or handedness on a plane surface. This means that one cannot distinguish clockwise from counter-clockwise on a

Klein bottle. Diagrams of the Klein bottle in 3D space, the handedness changes as a point passed through the self-intersecting neck while traversing from the "outside" of the bottle to the "inside". Again this is a strange property for the Klein bottle because it has only a one-sided surface the indicates Elliptical geometry of the atom boundary and inside.

While a Möbius strip can be embedded in a three-dimensional Euclidean Space $\mathbf{R}^3$, the Klein bottle cannot. But it can be embedded in $\mathbf{R}^4$, 4D spacetime one more dimension is needed to complete the twisty bridge without a discontinuity at the "handle" entry point. And the one-time dimension is sufficient because that twist in time changes spacetime into another kind of spacetime and also provides curvature, that property of spacetime that creates real physical fields and forces.

A possible model depiction of what this would physically look like below is shown in a Kleinish topological structure morphed into a one-sided sphere. A physical 3D bottle topological configuration of this kind, a shrunken umbilical and bottle orifice, goes a long way to provide the requirement for atomic poles and angular momentum about a pole axis. This is conducted in compliance with Topological rules, morphing the shape of the typical mug-shaped Klein bottle into one more closely resembling a sphere, as we need a vibrating sphere model to yield the "electron structure". So our first step is to reduce the handle to much flatter and shorter keeping the twist and dimensional transition but minimizing its physical dimensions. Morphing this topological shape without violating the topological and other mathematical rules provides us with a mathematical model, matching all of the attributes, properties, and characteristics of an atom, into a physical model.

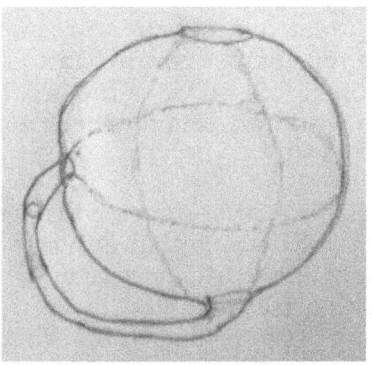

Most of us have seen a Mobius strip, where a one-sided surface twists into a one-sided surface. A Klein bottle has two Mobius strips inside of a bottle whose outside surface turns into its inside surface turns into its outside surface. The Klein bottle is a one-sided closed surface with a topological property called nonorientability which has left-handedness indistinguishable from right-handedness on the same surface but on different sides of the time dimension. Scootching along the "outside" bottle surface in 3-dimensional space we would be in a left-handed surface but once through the self-surface intersecting umbilical or neck segment, we would find ourselves on a right-handed surface. If this were just a spherical surface, we could not have the clockwise and the counter-clockwise topological property and so the sphere is called an orientable surface. Right-handedness and left-handedness become important later to show that the geometric properties of this same surface can produce a 2 pole entity, for magnetic force, and also a 2 opposing curvature surfaces on the same surface for the "outside" to be an attracting curvature and the "inside" to be a repelling curvature. In translation to another spacetime, this provides the architecture of EM, complete with a two-pole magnetic construction.

Projecting to higher dimensions, a 4D Klein bottle describes the boundary from non-Euclidean Hyperbolic spacetime to atomic spacetime, EGSFT. Atom boundaries are essentially 4D Klein bottles in the average shape of a 3-Sphere comprised of a vibration sphere modeshapes at natural frequencies of the sphere harmonics. Distinguished from the 3D bottle created by the great mathematician Felix Klein, the 4D Klein bottle throat is in the 3D spatial center which acts as a time gate, for frequency – monotonic time transition. This is because time cannot enter the bottle from the space boundary and must make an orthogonal twist to enter or leave the atomic 3-sphere boundary. Just like a 3D object can enter a 2D object without penetrating its 2D perimeter, likewise, the atom is a 3D sphere of space where time inverts from circulating to continuously monotonically increasing upon transformation from an atom's center to outside the atom. The frequency-time, in this 4D Klein bottle, can be treated as a frequency, acting on the space within the atom, whose spatial dynamic boundary, giving it space and spacetime curvature which mathematically dictates the physical properties of the atom that we measure, but yet do not treat as a vibrating sphere curvature boundary between the two spacetimes. See the figure below for a rough rendering of the Klein bottle in Minkowski space.

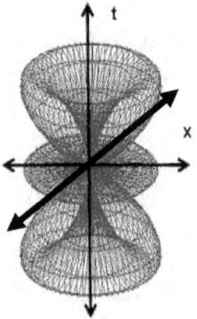

A Possible 4D Klein Bottle in spacetime

As the Klein bottle shows, time changes from the macro spacetime monotonic to a circulating time loop inside. Thus spacetime curvature transforms from outside the atomic boundary sphere by a Mobius transformation or a mathematical inversion operation to inside the atomic boundary.

Thus, by analogy and some reasoning, the straight-arrow of time seeking infinity at the big band under the formation of spacetime shear, found infinity faster in a 4D Klein bottle time vortex, much like eddies in a stream near the shore. Time found itself wrapped and trapped traversing the bottle, forming another kind of spacetime in another kind of perfectly valid geometry. The simplest case of such a formation will be a Hydrogen atom, forming the bulk of matter in the universe. Since observation has proved that this confinement is reversible and that in reversing, i.e. time exits from circular or frequency to monotonic in Hyperbolic geometry space, the resulting space dynamic that we see comes in the form of EM.

How can I wrap my head around this strange concept of physically cycling time?

The best argument that can be made came from Einstein's good friend Gödel and his solution to Einstein's EFE having looping time with "causally-connected events closed in itself". The solution from the Gödel solutions suggested that in a certain series of events there was no way of knowing which happened "earlier" or "later" than another event in a series of events.

The lack of "time's arrow" or time only going one way like an arrow, forward, inspired Einstein to write regarding Gödel's solution, "It will be interesting to weigh whether these are not to be **excluded** on physical grounds." It was deemed, because of a "Cauchy surface" constraint, that Gödel's rotating universe space could not be "globally nonhyperbolic", and would thus have to be bounded. While from a cosmological standpoint Einstein's comment makes perfect sense in our physically known universe, from the standpoint of modeling the inside spacetime of an atom with a bound physical disparate spacetime, this makes also perfect physical sense. So don't despair if you cannot wrap your head around frequency-time, Einstein couldn't either.

However, what is already widely excepted, physicists teaching curled spatial dimensions as additional higher dimensions is pretty unsettling as well. These extra spatial dimensions are simply additional rows and columns populating an ever-growing Metric Tensor. This method was first proposed by Kaluza to "unify" Gravity with EM, something Einstein was working on at the time. This idea was initially thought preposterous even by physicists, though Einstein wrote that he believed it was a mathematical anomaly with no physical interpretation. A decade later, Felix Klein proposed a hose physical model. From a distance, a hose looks like a one-dimensional object, a line. But closer inspection reveals a tubular configuration with thickness and dimensions curled up inside. The thickness radial dimensions inside a straight one-dimensional line to add 3 more curled dimensions per each macro x,y,z spatial dimension. Klein was on the right track to a physical model but stopped short of a current standard of "plausible reality reasonableness."

Prior to that, a theoretical physicist named Gunnar Nordstrom introduced an additional space dimension in his model, which was the first of the extra-dimensional theories. These discoveries started a series of string theories.

Regarding prior ruminations of cyclical time, in the literature, we have Alice in Wonderland by Lewis Carroll. Carroll conjures reference to Time-warp in the form of the March Hare, Mad Hatter, and the Door-Mouse, all are trapped in a perpetual time when tea is being forever served. Life is one long tea-party as it's always tea-time, six o'clock. While the atom is a little more confined space, the physical idea is similar and at least conceptually compelling, well maybe to regimented habitual tea drinkers.

An argument could be made that the other matter and atom models are way crazier and so Chronature is Occam Razor more plausible by default. More recently, the TOE's bran model over the Superstring Theory's string, it was noted that by curling the bran into a cylinder and stepping away, one can obtain a string. In comparison to these current models, curled time should not be a big stretch. In mathematics, mapping from time to frequency and frequency to time domains has been known for hundreds of years, appreciated, and used in many applications. From Laplace and Fourier transforms, frequency is the physical mapping from the time domain, monotonically increasing time-dependent functions mapped to the frequency domain. The units tell us frequency is the reciprocal of time and a viable alternative depending on the application.

Going back to the alternative theory, QM, one does not have a time dimension, it's all a matter of probability. Einstein found QM philosophically unsatisfying and thought that it was wrong. In his 1935 paper on entanglement, Einstein wrote: "One is thus led to conclude that the·description of reality as given by a wave function is **not complete.**"

Einstein further believed that randomness is a reflection of our ignorance of some fundamental property of reality. He was a "determinist" and believed that if we see some physical thing happening, then there are clear reasons why it happens and that if we know causes A B C, that something should be repeatable. *"God does not throw dice"*, in his oft-quoted phrase. Einstein never accepted the uncertainty principle, because it states that in some cases we *cannot* make definite predictions. Without good relevant predictions, a model suffers along with the model proponents.

Einstein believed that if we can't make a definite prediction, then it is because we don't have the necessary and sufficient information about the situation, or an accurate theoretical model that allows our models to make valid predictions. And that says it all right there, if your model cannot predict what one can see and or measure, then its unreliability and its use to predict is suspect. After all, a bad model can give is the correct answer sometimes and not others, like a broken clock.

Along with Podolsky and Rosen, Einstein (EPR) eventually wrote a paper with what they thought was a thought experiment whose results were so outlandish that they couldn't possibly be true. His purpose was to offer a QM model solution that proved QM could not possibly be a true or complete description of reality because following QM through produces or yields an "impossible" prediction. QMers claim that they've done experiments in a multitude of different ways and quantum mechanics is indeed correct. However, since QM is not a model of physical reality and only probabilistic, QMer claims that EPR yields real-world predictable results is a contradiction itself since it leaves out time. Gaps and loopholes in the QM leave reality inconclusive or incomplete, with a probability of 1 that its usefulness is dubious without some lucky right answer despite the model itself being bad.

As an aside from an idea evolving through its overzealous accolades I have this to offer. Upon being asked why he "did it" Madoff's answer was something along the lines of that he didn't' mean perpetuate the fraud but he needed to square the books from a previous night's account cooking, which he planned to rectify the next day. His corruption started small and he had all the intention to repay investors with no one the wiser. But the borrowing from the shorted accounts to cover up requirements from borrowed accounts kept growing and so Madoff kept increasing the "borrowing" to protect the lie he had concocted to serve reality, here the duped investors.

QM has always struck me as having evolved in that same fashion. Accidentally obtaining the solutions for a vibrating sphere to accurately model electron orbital structure provided "proof" and hope that the Bohr model was good enough. That did not change the fact that QM theory was incomplete, not a physical model and unreal, but for the proponent's need to perpetuate the theory their explanations would get weirder and stranger until it is what we have today, incomplete and completely unmanageably "complex" for anything outside of a hydrogen atom and too crazy to tell inside of a hydrogen atom. Under the QM model, human consciousness becomes a causal component in physical phenomena because possible "states" of physical reality are fluid until an observer makes them all "collapse" into one state that is observed.

Because QM does predict some very basic things like electron shell structure but does not allow for complete determinism, we should be attempting to figure out a better model instead of pushing to expand an incomplete or partial theory. The Chronature model is not probabilistic or non-deterministic and is capable of some physical confirming observations, explaining some unexplained, and predicting measurements. So frequency-time producing an alternate spacetime may be easier to swallow than QM or String Theory.

Since the atom is a 3-sphere, time rate C is constant and spheres being similar, how do we get so many different elements?

The short answer is because the proton nucleon is asymmetric and is constantly bouncing around in the atom, impinging on the atom wall. The more of these the farther they push the atom wall.

The atomic element number depends on the number of nucleons and in particular protons in the "nucleus" or center of the atom. The nucleons themselves are smaller, tighter spacetime 3-spheres, 10^{-5} times smaller in Euclidean space than the atom boundary as calculated once outside the atom in HGSMT spacetime. This is because, inside the Elliptical non-Euclidian Geometry and Frequency-time atom, measurements would be vastly different. The spatial Elliptical Geometry dimensions inside the atom will call for different dimensions for the nucleons, themselves comprising yet another type of spacetime. Each addition of such a proton nucleon, adds harmonics to the 3D sphere curvature vibrating boundary of the 3-sphere atom boundary because the proton nucleon has eccentricity which acts against the boundary producing a spherical harmonic oscillator. Thus the atom's spacetime boundaries are very dynamic due to the movement of the asymmetrically shaped protons. The neutron nucleon, however, is quite symmetric and adds no geometric perturbation oscillation to the inside of the atom wall spacetime boundary vibration. The eccentricity of the proton, as we shall see, is what gives the proton "charge", the same as the eccentricity of the much larger 3 dimensional spherical harmonic 3-space dimensional lobe-shaped electron. Moreover, because time is circulating inside, the atom boundary is fluctuating with time outside from time inside, hence space curvature is dynamic and the atom exhibits "chemical", "magnetic" and "electrical" properties, based on the curvature of the atom boundary dynamics. The dynamic curvature atom boundary between the two spacetimes follows fluctuation in accordance to a 3D spherical vibration, with harmonics that depend on the number of protons forcing oscillations inside the atom. The addition of harmonics to the atom boundary, what we will call electrons, produces contributing curvature to the atom 3-sphere's fluctuating in time membrane boundary, changing the overall curvature of an atom to contain the additional harmonic in the atom boundary. It is the atomic boundary curvature that then gives the atom the electrical, chemical, Van der Walls, and other properties. Analogous to the Fourier series of sinusoid harmonics comprising any continuous graphical locus, 3-dimensional spatial pod and sphere-like modeshapes are spatial harmonic solutions. Together with the temporal harmonics these sum to define the curved outer surface curvature of any element atom in space and spacetime.

    The local curvature of the sphere boundary gives each element its properties and characteristics in a local sense but with the time of contact included. For example, an element whose atomic boundary curvature is symmetric, as in one of the

Noble gases, as its rotational symmetry will display an equal attraction from all sides will hence be relatively inert, attracted to all boundaries equally and none in particular. Alternatively, an element with high eccentricity or surface curvature difference will have a high attraction to another atom, at the surface point of the highest curvature. Thus the periodic chart will organize elements based on these properties of elements into "families". These come about naturally as a consequence of spherical harmonics which produce symmetries and asymmetries in the atom wave boundary, or space curvature between the spacetime inside the atom and the spacetime outside the atom.

If all atoms are all just another kind of spacetime, why then don't they just merge, after all, space is just nothing or the absence of anything?

This would be true if time were the same inside as well as outside of an atom, as the spatial dimensions could then be transformed to spherical geometry and back. Since time inside the atom is different from the time outside the atom, any object having an outside atom time will be attracted to the atom boundary curvature, by virtue of the atomic sphere's outside curvature being negative, ie as identified defined and described by GR and shown in the figure above, the attraction of matter in moving but bent spacetime. GR holds that time runs slower near bodies exhibiting spacetime warp, see figure above, going to zero at a black hole event horizon, the densest matter that we know. In addition, simply adding different times together would be like adding the future to the past, they simply cannot exist, a violation of causality and strictly prohibited by our physical reality spacetime. It would be like jamming one position of space into another; uniqueness in spacetime curvature is preserved.

    Once atom boundaries touch, they don't just merge into a common spacetime unless they have sufficient attracting time and space curvature to overcome the repelling spacetime curvature inside the atomic membrane, and can remain stable in the final merge state. Also, there is something called "time censorship" whereby measurement or observation of the event may not be possible across disparate spacetimes.

    Except at absolute zero, condensed matter, atoms have vibration, rotation, and translation, changes in space and time. This indicates that spacetime curvatures are highly dynamic. Thus the vibration, rotation, and translation from the colliding spacetime boundaries will produce partial topological morphing of 4folds which must comply with the mathematics of 4 fold motions bearing in mind that positive curvature spacetime inside the atomic membrane or boundary will repel with encroaching spacetime curvature.

So then what happens to time beyond the event horizon and inside the steeper spatial gradient?

    Moving back to inside the atom, time is inverted to frequency and by direct analogy by vastly different scales. We could just as well model that time at the event horizon of a black hole and also the atomic boundary does not become infinitely slow but transforms into frequency or periodic time within that atom sphere boundary or black hole event horizon. This also solves the problem of singularities for space and time in a black hole on the macrophysical scale. A black hole acts like a giant atom, where black hole curvature of spacetime is a sphere with circulating time inside, all of which reached a kind of "critical mass" to maintain the positive curvature space inside the event horizon, forcing any swallowed matter toward the hole center much like the "strong force" does in an atom. Hence, by analogy, the black hole should be capable of Hawking escaping EM radiation at discrete frequencies and on devastatingly large proportions, or "gravity waves".

Since frequency-time inside averages to zero time, even though it has a value other than zero, any monotonic time object caught at the event horizon is inverted and the spatial geometry thus changed to Elliptical, inside the event horizon. And space lines go to infinity inside a black hole, but they are wrapped into an elliptical geometry geodesic following sphere. Hence there is no singularity in physical space inside the black hole, only a Mobius or inversion transformation into another spacetime.

When one atom enters another atom's boundary, the entering and entered atom have two different time and spatial positions. The disparate spacetime curvatures will exert disparate accelerations, forces, on each other which in most cases will push them apart. These, depending on the velocity of entry, cannot violate the spacetime uniqueness rule, and must then morph their existing boundary times and positions, yielding changed velocities and hence trajectories in the interaction. A change in velocity of the "impinging" entity indicates that acceleration occurred, as we expect from the curved spacetime acceleration, and this provides the appearance of forces acting. Thus two regions of curved spacetime appear "solid" because they bounce off each other like billiard balls. But "solid" is only an observation of atom spacetimes interacting and resulting in our current perception of the interaction between "solids".

How come we cannot create matter even though we know from $E=mc^2$ that they are equivalent?

Scientists cannot create matter, it is just here and there and we can use it. We can convert other matter or matter into energy, fission and fusion are examples, but we cannot create them. We can change one type of matter to another, but we cannot create it from just energy, even though the Einstein equation shows that there is that equivalence. Even the creation of anti-matter requires matter.

Physicists may argue that they create matter at accelerators daily. This is done by smashing together two or more particles which results in producing several particles that have greater mass than the two initial particle reactant mass. Hence they regard this as the opposite of "mass defect" where the product particle(s) is less massive than the sum of the reactant's mass and "energy is released" as in a nuclear device. But where did this extra "missing mass" come from? "Nucleon binding energy" whatever that is. Having a name for something doesn't mean that's it's known or that's a reasonable explanation. Bound nucleons didn't just give up some "mass" to lighten the atomic load for fission or fusion. Protons and neutrons don't change weight or mass. Mass is matter in the form of spacetime curvature. The fact that some curvature decided it was going to transfer into some other kind of curvature has nothing to do with the creation of new matter. The "missing" mass defect is a reduction in spacetime curvature inside the product fragment atoms. So where the extra matter came from in the "more/less mass in products" is the same answer for the opposite effect of "mass defect" where reaction absorbed "energy" is necessary for the formation of product particles whose total mass is greater than the total sum of the product's mass. That reformulated "energy into mass" must have come from the addition of curvature. But energy did not create matter, matter was instead used to create other matter plus or minus "missing mass" in the form of energy or spacetime curvature. "Energy" plus matter formed different matter. As we shall see, "energy" is just an artifact of spacetime curvature, outside and inside of the atom. Without combining with the existing matter, no energy can be created. The reason being, that flat spacetime has no curvature, therefore no acceleration, and therefore no forces can materialize without some matter present. No forces mean no energy. No energy means no matter and so forth, wherever there is energy, there will exist spacetime curvature in some form, mostly matter and EM.

Cosmologists tell us that somewhere in the great big universe, matter is being created. We posit this is happening at the edge of the Minkowski cone universe as shown in the figure above. The Einstein Cosmological Constant helps model this for an expanding universe. So if matter is being created at the edge of the universe, we are still guessing on what, how, and why this is being accomplished. We can only speculate, as the forces at play at the Big Bang, pre-matter formation, when the laws of geometry had not yet been formulated, and some kind of spacetime shock wave continues to create spacetime turbulence forming spacetime "bubbles" ie spacetime vortexes or curvature in the various forms of matter, in its wake. The edge of the universe, the gigantic spacetime shock wave, is still expanding, and matter is being created.

Why the Matter-Energy equivalence?

We have been taught that matter and energy have been defined to some level by physicists and that they are equivalent and brought to you by $E=mc^2$. The current model that matter and energy are equivalent to a physical entity is tenuous at best, based mostly on the "mass defect" from fission and fusion phenomena. But in the Chronature model of matter, without having to define particles, matter and energy are both aspects and artifacts of the same things, space and spacetime curvature. That is why matter-energy equivalence; they are the same thing, spacetime curvature in different shapes and forms. We proffer that the currently accepted three conservation laws dealing with matter concerning continuity, momentum, and energy derived from the one conservation law, and that is in the dynamics and transitions of spacetime curvature from one spacetime to another, spacetime curvature is conserved. It sounds simple and in that way complies with Einstein's dictum that a good and elegant theory "must be simple but no simpler."

As a start, an atom is a moving, vibrating, and rotating completely enclosed positive curvature vortex of spacetime within a spacetime curvature boundary. Kinetic energy is the space and spacetime curvature that causes acceleration forces and also transitions between the different spacetime geometries of hyperbolic HGSMT and elliptic EGSFT spacetimes. As curvature from within an atom boundary, curved with time, is transformed into the curvature of another kind of space, the space of infinitely monotonically increasing time, the positive curvature must relax from tightly bound curved space with frequency-time to straight, geodesics, relaxed sheets of space with monotonically increasing time. The transition takes the form of EM. Thus electromagnetism is spacetime curvature transforming from positive curvature space-frequency-time, EGSFT inside an atom, to negative curvature space monotonically increasing time HGSMT. As we show later, an EM wave will resemble a train of sphere-like one-sided surface, because the frequency-time must map to monotonic time and the space curvature, whereby spherical harmonic modeshapes from inside the atom, translate that to rotation along the pole axis of multiple space harmonic petal-shaped curvature, the spatial dimensions adding frequency or circularity to each spatial dimension.

Why do the nucleons congregate in an atom's center against the electrical repulsively charged protons, also called the strong force?

It is generally accepted that the atom center is a potential well, somehow confining the positive charged mutually repelling protons, trapped in the atom center in a "potential well." This is AKA the Strong Force, SF, but its physical nature or mechanism is not understood.

What is known and accepted is that the neutrons are neutral of charge but slightly more massive than protons, yet they are confined by the Strong Force just the same. Since they have no charge, the SF is not electrical. But what precisely is the "potential well" that keeps the like charge protons from repelling themselves out of the nucleus?

Spoiler alert, SF is repulsive gravity. How could that be, since we all know that gravity is only been observed to be an attractive force, frequently used by the massive bodies of matter. Well, that's in accordance with GR, where spacetime has negative curvature. This happened where massive bodies are concerned but inside the atom, we find positive spacetime curvature, a curvature sign change that creates the opposite of attractive, a repulsive gravity.

We postulated that inside the atom boundary, EGSFT curvature spacetime prevails. That is inside we have Elliptical non-Euclidean Geometry metric and outside we have Hyperbolic non-Euclidean geometry metric. In a 3-sphere, we have a negative curvature spacetime inside, which acts opposite to the positive curvature spacetime curvature outside or reverse the convention, that is local reference frame objects compel objects away from the boundary instead of attracting it, and pushing inwards towards the 3-sphere, atom center. That's correct, a repulsive gravity force, not attractive much like the "north" pole of a magnet, only much stronger. This is a strange notion knowing what we do about gravity in general and as shown in FIG. 1.1. So it meets the "crazy enough" standard that the Strong Force (SF) is a gravitational repelling field, since spacetime curvature is negative inside of an atom, the opposite of the attractive gravity force that we experience from positive spacetime curvature, and many orders of magnitude stronger because spacetime curvature is so much more steeply curved, much smaller radii of curvature, inside of an atom than on the circumference of the earth. This is shown in FIG. 1.2 below, and includes FIG. 1.1 to better illustrate the gravity force direction on each side of the atom boundary.

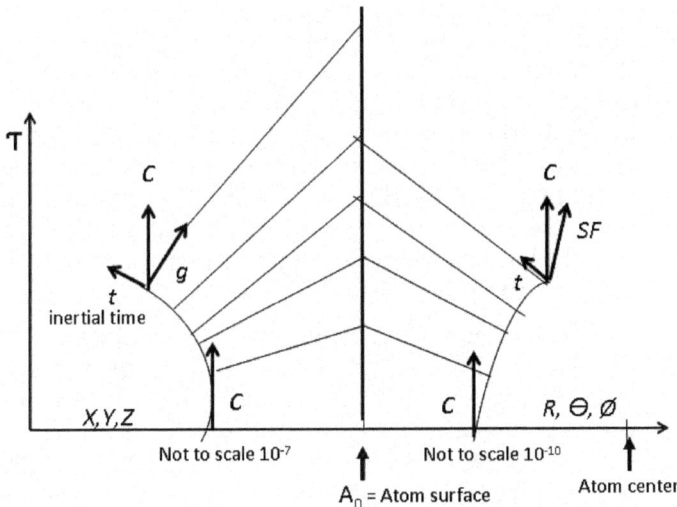

FIG 1.2

FIG. 1.2 is a modified Minkowski diagram, we can see with merely the progression of time alone, the curvature inside of the atom gives rise to an atom boundary with repelling gravity, SF, inward. This is known as the Strong Force, SF, but here shown with the spacetime curvature as it exists inside the atom, it is analogous but opposite the attractive gravity g developed, again, solely due to time progression and curved spacetime. A "gravitational collapse" may be a better analogy as the mathematics would be similar, except and a much much smaller arena.

    Again, because it is critical for a unified field theory model of physical reality, inside the atom the SF is gravity opposite in curvature sign, giving a repelling force from the atomic boundary and toward the atom center, where the nucleons are compelled to reside, without the need for "gluons" and through a centrally directing curvature repelling gravity alone, keeping all of the nucleons happily drifting around at the atom center or nucleus despite the "positive charge" protons repelling one another by "electrical" forces in their attempts to escape. Note that the curvature and distance within the atom are not to scale but that the SF vector would be sizably larger than the atomic forces without just due to the tremendous increase in curvature of the much smaller nucleons.

    Oppositely, the negative curvature and monotonic time outside the atom boundary curvature form an attractive force to objects outside the atom through spacetime adjacent to matter from the warp. Nucleons, inside the atom, will thus congregate or be repelled from the atomic boundary and towards the atom center but will wander slightly because of the proton eccentricity or non- SU(3) symmetry. Put another way, a spatially stationary reference frame will, under the influence of frequency-time alone, drift towards the atom center at the speed of time C (as shown in FIG. 1.2). Hence the nucleus "potential well" is simply an attribute of space and spacetime inside the EGSFT conforming atomic sphere boundary, where the gravity direction is reversed, resulting in a very strong repelling gravity to the atom's center, known as the Strong Force. Thus curved spacetime alone can create another of the Universe's 4 forces.

What is the physical structure of magnetism from the standpoint of spacetime curvature and how is it physically manifested by spacetime curvature acting alone?

The physical Chronature model also provides a unified field theory and must account for physically how and why the magnetic force manifests physically in the way that it does from spacetime curvature and geometry alone. One of the prime identifying characteristics of magnetism is the two opposite valence poles, called the north and the south. The north pole is by convention the repelling magnetic flux end and the south is the attracting flux end.

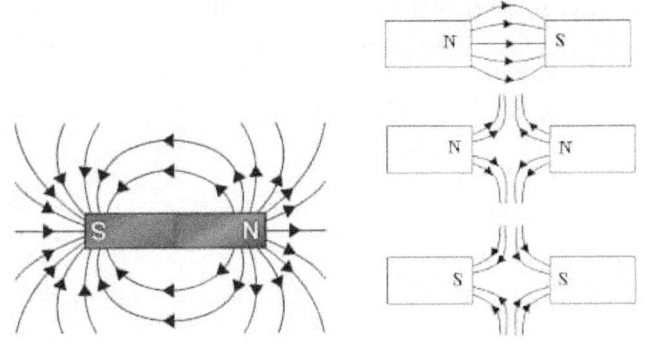

This all comes about through 4D spacetime structures which contain both negative curvature spacetime as well as equal but opposite positive curvature spacetime. And they travel together to produce the dipole repel-attract magnetic field in a single atom shown just below or through a group of atoms in a lattice structure or "domain" as shown just above in a bar magnet.

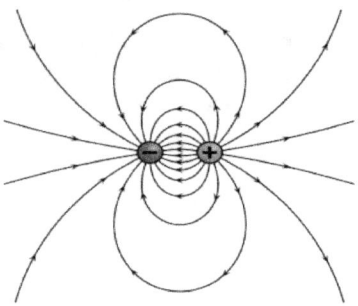

In a way, they are like combined strong force and gravity, a result of positive, or negative depending on the convention, spacetime curvature, and a strong opposite attractive gravity force, negative, or positive with convention, spacetime curvature. The Minkowski diagram below illustrates how and why the magnetic force manifests in the way that it does, with the repel-attract "north" and "south" poles.

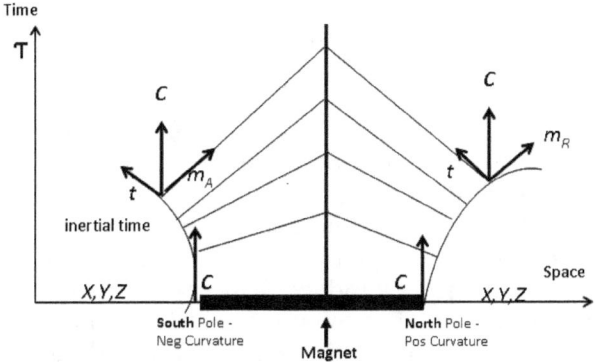

Please note the m_A is the attractive force south pole develops solely through the passage of time in a warped spacetime, and m_R is the repulsive magnetic force on the north pole developed solely with the passage of time in a warped spacetime. As with GR, where lines of spacetime are longer near matter, a spacetime warp or curvature is created. The South magnetic pole has longer negative curvature spacetime lines and hence forms an attractive force, like gravity as we know it, attractive. This "magnetic" field is formed by virtue of time progressing alone in the 4D manifold of curved spacetime. As time vector C progresses in the curvature of spacetime, a magical force, like gravity, manifests as simply a component of the spacetime grid by virtue of the inertial local reference frame of an object in the field rotating to comply with the spacetime continuum local axis. As the diagram just above shows, the magnetic north pole, having a longer positive spacetime curvature, acts in the opposite by repelling objects away from the atom boundary because its spacetime is curved in the positive direction. "Magnetic" force is developed also merely through the progression of time, and in the same direction as the Strong Force, repulsion.

If one were to cut any magnet into pieces, each piece would automatically develop a north, repulsive, and south, attractive, pole. Since this occurs down to the atomic level, an atom should be able to develop magnetic forces as well. And we can show that this does indeed occur through spacetime geometry alone. The mystery is in how this happens, but the north and south poles maintain their valences.

In 1958, Smale astonished the mathematical world with proof of a sphere eversion. Well, not quite an eversion because we will start from only a one-sided surface sphere, but we will make a sphere that has two equal and opposite symmetric curvatures on the same side. This would be somewhat akin to differential topological eversion of a sphere, but instead of moving the inside sphere surface topologically to the outside, our sphere will move the inside curvature to the outside curvature on the same one-sided morphed sphere. We will create a one-sided surface sphere that has two equal and opposite curvatures. We will still require a vibrating sphere or spacetime entity to model the physical harmonics of an atom but we must further our model to provide a spacetime entity that will have equal and opposite surface curvatures. This is illustrated below in the step-wise topological transition from the Klein bottle to an F-sphere.

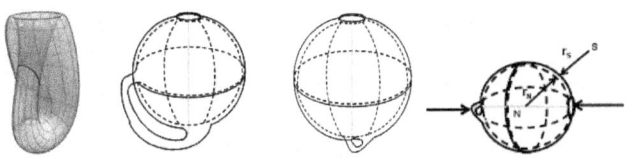

FIG. 1 From Klein bottle to F-Sphere

Please note that this F-Sphere configuration depicting a relatively small handle at the bottom can be much smaller, morphing residue from the much more handle-looking Klein bottle. The Klein bottle handle virtually disappears in the F-Sphere by morphing into a tiny dimensional twisting appendage to create the sphere essentially a one-sided surface with an attractive pole, and a repulsive pole with spin axis orientation in space and time. Moreover, these little one-sided surface twists and turns become more important in explaining physical atomic characteristics such as spin, electron states, parity, and bonding. Magnetic dipole moments and magnets arise from these types of spacetime curvature configurations and their combinations and orientations which accumulate into what is currently called "magnetic domains."

An interesting attribute of an F-sphere, a sphere-like entity with one point missing at the north pole, is that it is diffeomorphic from a Klein bottle and can be mapped to a flat plane. This becomes important when we try to calculate spacetime curvature values; the metric between the projections is inherited in the F-Sphere. The topological properties becomes important as we mathematically traverse the surface in spacetime and find it to be equal to zero. Restating the obvious but necessary, the changes from a conventional state to a new topological state change the spacetime curvature giving the changed entity new physical properties.

With this physical configuration of spacetime surface, an F-sphere offers is a topological model of the atom that has long straight lines, warping spacetime, negative curvature spacetime on one pole, and equal amounts and degrees of positive curvature spacetime on the pole of a klein bottle morphed 3-sphere. R_S, R_N- below:

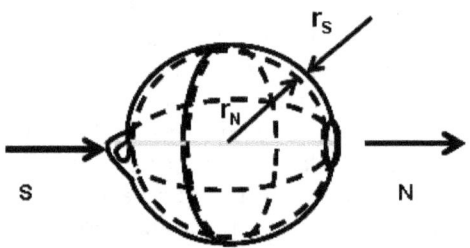

The morphed one-sided surface 4D Klein bottle atom configuration is shown above illustrates attracting negative curvature, "outside" the sphere, as calculated from radius of curvature r_S of the atom boundary, simultaneously creating longer lines to "outside" space, completely equal to the spacetime on the "inside", negative curvature from the radius of curvature r_N, of the atom bottle. Hence the "south" pole develops the attractive force and the "north" pole develops an equal repelling force of a two-pole magnet or dipole. So in the spherical atomic configuration, the "inside" spacetime boundary creates a force compelling objects toward the center.

Moreover, this through nothing more than changes in flipping orientation of the normal in compliance with a one-sided surface geometry of space and time. One might wonder why this doesn't quite physically look like an earth with a "north" and "south" pole on opposite sides of the sphere geometry. The depiction in the figure above offers a 4D surface configuration as an explanation of how the geometry of spacetime can be among other things configured to manifest a two-pole magnetic field dipole, from an atom's boundary through the geometry of spacetime curvature alone. Since there is only one side to the surface, time orientation wrt the surface will produce the dipole field. So, matter is comprised of many atoms aligned synchronously and spatially aligns additively to define a magnetic field. In this context, a magnetic field would be more alternatively called an attractive-repulsive push/pull gravity field or a two-pole magnetic field from spacetime curvature. It would be like a hill and valley that always travel together.

    A nuclear physicist once was asked what his motivation for splitting an atom was. He quickly responded, "well somebody may just want a half". It would appear the monopolists are looking for just that half. But we can see here from purely geometric models that the magnetic field comes only in pairs because a one-sided atom surface in twisted spacetime provides that both the curvature of the inside and outside are equal and part of the same topological entity having equal opposing surface normals. Hence there is no "monopole" as that would mean we would have to surgically divide the atom in half, exactly, so that the curvatures would be on separate entities, half atoms.

Because an atom has two poles, equal and opposite symmetric negative and positive curvature, they can form a submicrometric system that presents spontaneous magnetic order at zero applied magnetic field, "residual magnetism" and through the application of moving electric field. We will discuss E-fields on more dept below, but suffice it to say here that E-fields are acceleration fields from spacetime curvature which are integral with a curved surface providing a B-Field, magnetic field, also acceleration fields arising from spacetime curvature.

The small size of atom magnetic property does not necessarily respond to the formation of typical magnetic "domains" and the magnetization dynamics of atoms at low temperatures, are typically viewed as "quantum phenomena" or" spin tunneling". At larger temperatures, the "magnetization" undergoes random thermal fluctuations or what is known as superparamagnetism. The point here is that Chronature atomic spacetime curvature models explain and even make obvious, many physical phenomena without resorting to a probabilistic and non-physical or intuitive model.

How many sides does a surface have?

To the hardened mathematicians and theoretical physicists, the F-Sphere will look "wrong" because of the mathematical property of orientability. Yet the Klien bottle has been known and studied for over a century. Orientation and one-sidedness have to do with fancy things like rotation, circularity, handedness, and other arcane mathematical properties of topological surfaces.

There are generally two types of surfaces, differing in the way in which they are situated in the ambient or entity containing space. For example, a cylinder is a two-sided surface, while a Mobius strip is a one-sided surface. A characteristic distinction between these surfaces is that the boundary of the cylinder consists of two curves, while the boundary of the Möbius strip is a single curve. Among the closed surfaces the sphere and the torus are two-sided, while the Klein bottle surface is one-sided.

Mathematicians[9] would characterize one- and two-sided surfaces in two types of manifolds differing in the way in which they are embedded in the ambient space that is a space of dimension higher by at least one. If you let them, mathematicians will tell you that two-sidedness and one-sidedness are related to orientability and non-orientability, but unlike these, are not intrinsic properties of the surface and depend on the ambient space. For the mathematician for example, there exist orientable two-sided surfaces: $S2 \subset S3$, $T2 \subset R3$; non-orientable two-sided surfaces: $RP2 \times O \subset RP2 \times S1$; orientable one-sided surfaces: $T2 = S1 \times S1 \subset RP2 \times S1$; non-orientable one-sided surfaces: $RP2 \subset RP3$ (here $S2S2$ is the sphere, T2 the torus, RP2 the projective plane, RP3 projective space, and S1 a disorienting path on RP2). In an orientable space (e.g. in Rn) a hypersurface is orientable if and only if it is two-sided.

9

https://www.encyclopediaofmath.org/index.php/One-sided_and_two-sided_surfaces

The problem is the consistency or inconsistency of normal surface vectors, as this becomes important in modeling curved spacetime to explain physical phenomena. Moving along a smooth surface embedded in some space independently of the choice of the curve, and where it remains the same normal to the surface, the surface is called two-sided; in the opposite case, it is called one-sided. And for the F-Sphere surface, in traversing the surface normal is flipped to exactly the opposite directions, 180°, almost magically leaving us a one-sided spacetime surface with two equal and opposite normals, equal "poles" and spacetime curvatures on the same surface.

What is Magnetism?

Magnetism is yet another type of spacetime curvature resulting in a physical manifestation of force and fields. Science currently describes and models them as magnetic fields. There are several types of magnetic phenomena, currently labeled ferromagnetic, diamagnetic, paramagnetic, antiferromagnetic, or combinations of these. A magnet's most elementary spacetime configuration is a dipole, having equal and opposite spacetime curvature on an F-sphere. The two opposite spacetime curvatures give rise to an attractive surface and a repulsive surface on the same one-sided surface of an atom as described in the F-sphere Chronature model atom. The different magnetic field strengths or properties result as a consequence of the alignment of neighboring lattice F-sphere or atoms and their movement or stability within a lattice, giving us some substances with each magnetic property as a function of temperature, pressure, and field strength. Electricity, the transmission of electrons aka plane front propagating atom boundary vibrations between neighbors in a lattice can also align F-sphere dipoles which create magnetic fields of various types and strengths.

As with all such spacetime curvature-inducing sources, their "strength" always decreases with distance from a magnetic source which depends on source boundary, configuration, alignment, and stability. Various configurations of magnetic moments, atomic lattice, magnetic dipole material properties, and electricity can result in complicated magnetic fields.

This brings us to magnetic and non-magnetic materials. What are they and why are they? Magnetic materials have atoms and compounds that are aligned on a macro scale, north to south poles arranged synchronistical and positional to amplify and extend the reaches of a field additively. Heat and shock loads will ordinarily rattle to randomize individual atom or molecule pole directions, misaligning the atoms and diminishing the magnetic field strength as alignment decreases. Super magnetic materials configure the atom lattice in a way such that the random vibration-translation of the atom neighbors does not randomize any atom poles out of domain alignment, moving added energy into simple symmetry invariant atom rotation of the atoms in the lattice. When this occurs at much lower temperatures, less atomic lattice transitions to reduce the amount of atomic vibration-translation that can occur, acceleration channels in the lattice open and we see properties like superconduction and superfluidity.

Why is gravity such a weak force compared to the other 3 known forces?

Some models explain this by "leakage", that is the force of gravity is dissipated somehow in the higher dimensions proclaimed in String Theory or TOE. But if you are even semi-literate you will know that like Santa and the Easter bunny, no extra dimensions have ever been detected. So it is much easier to accept that the force of gravity, in accordance with GR, is just spacetime warpage on a large to very large scale in 4D. We proffer that EM, electric, magnetic and Strong Force, scale with dimensionality and not dimensions. The Weak Force, measured under most circumstances, turns out to be a consequence of the vibrating atomic boundary, in 4 dimensions, in a frequency-time and vibration-rotation-translation modeshape interaction with nucleons crossing the spacetime center in atomic space volume.

The point to be made here is gravity is **not** weaker; they are the same thing in a GUT fashion. As shown below, the spacetime curvature of an atom boundary relative to the spacetime curvature of an astronomical body is on the order of 10^{40} times larger for the atom boundary surface spacetime curvature, when the curvature of spacetime wrt space and spacetime wrt time is taken into account, consistent with GR and the EFE.

How is the Weak Force manifested in Geometry and what does the Weak Force physically represent?

The weak force comes about from the dynamics of the atom boundary in space and time about a common space and time center or origin. Although we generally model the origin of spacetime in Minkowski space as common, in physical reality this is not always the case. Remember that time is cyclical or harmonic "inside" of an atom, circulating about a reference time and spatial center. The deviation of the centers in space from the center in time allow for subatomic particles to "leak" through the physical atom boundary when the drifter subatomic enters the time and space co-inciding centers. How does this come about? Nucleons, protons and neutrons, drift about the atom 3-sphere volume center, confined in a region typically called the potential well built by the Strong Force, or repulsive gravity. The proton's eccentric shape causes a disturbance, or harmonic forcing function on the atom's boundary surface from the inside, giving way to the atom element excitation harmonic modeshapes or at a vibrating atom sphere with natural frequencies. The atom center deviation acts as a "time gate" for release or capture of transitioning space objects or traveling curvature. Depending on the number of nucleons, the size of the atom radius, number of protons, nucleons with "charge" character, these can be statistically released as the time gate at the sphere center open at times where time window and space modes are in resonance, and a nucleon drifting across the "time gate" and atom center area coinciding with the time window, can be released, as "spontaneous decay particle" emitting one or more nucleons and/or electrons from the boundary harmonic. Thus physically, the spatial center and the temporal center must resonate at the precise location of one of the wandering nucleons which then manifest as a spontaneous "decay" upon transformation from inside the atom to outside the atom, where the curvature is preserved. An analogous physical similarity can be found with the effect that the planets have on the Sun. We all know that the planets are gravitationally attracted and hence have orbits around the Sun. But the Sun is affected by the planets as well, causing the sun to wander around in the "barycenter." The entire solar system including the Sun orbits around a common center of mass called the barycenter. Due to the motion of the planets, the barycenter for our solar system is not constant. Sometimes it is inside the sun, and other times it is outside of the sun. The nucleons all wander around the geometric center which occasionally coincides with the temporal center, but the protons perturb the atom producing a forcing function for sphere vibration harmonics.

Why is String Theory an incorrect approach and where did String Theorist go "Not Even Wrong"?

String Theory is foundational upon higher spatial dimensions curled up inside the three "large" or macro well-accepted and well-known spatial dimensions, traditionally x y and z. The expansion of the metric tensor starting with Kaluza-Klein's 4th spatial dimension produced a unification of GR and EM. It was shown that extending the number of spatial dimensions in the metric tensor, led to a

Unified Field Theory, the holy grails of physics at the time. Thus it was reasonable to assume that by populating the metric tensor with more columns and rows, starting with gravity and EM, physicists would be able to explain and eventually show unification for the weak and strong forces as well and obtain the General Unified Field Theory (GUT) merely by defining more spatial dimensions into the metric tensor.

But this approach has not panned out for String theorists despite the best efforts by particle physicists diligently probing the structure of matter with their new tool, CERN. In distinction, by postulating spacetimes as building blocks, composite spacetimes, atoms residing in macro 4D space, built from their own separate and different 4D spacetime, content without the need for additional "curled up" spatial dimensions in our macro spacetime, create one big spacetime with 11 spatial dimensions. The Chonature model provides these "extra" dimensions from different spatial and temporal dimensions residing inside the atom and subatomic while creating differing kinds of matter. Much like electrons going backward in time forming what is called a different particle, positron, spacetime with positive curvature and frequency-time also form different particles and solely by spacetime curvature, a non-particle. Moreover, a negative frequency-time would produce an alternate temporal dimension in the metric tensor with possibly another type of "particle". Thus the metric tensor for a GUT can be modeled with 12 dimensions, four in HGSMT, four in EGSFT, and 3 spatial dimensions and a negative temporal frequency-time dimension for quarks and such.

For an 11 dimension, TOE the model, the Chronature atom model "curls" the time dimension with Riemannian geometry spatial dimensions into 3-spheres, or matter. Including additional dimensions of EGSFT spacetime, x, y, z, ω (r, θ, ψ, ω) and transitions from one spacetime to the other, all the additional dimensions in the metric tensor can be populated, but not with curled-up spatial dimensions as String and TOE Theorists hold, but with sets of spacetimes of three spatial and one temporal dimension sets. Viewing spacetime as co-existing interacting composite spacetimes, probing and populating the metric tensor in that fashion, also unifies the four known forces in the universe, without having curled spatial dimensions around macro spatial dimensions as proposed by Kaluza-Klein. Frequency-time inside of spherical spaces provides at least two separate and distinct spacetimes, populating the metric tensor with at least 8 rows-columns. More row columns can be represented with the transformations between spacetimes, EM, and the subatomic "particles" as well.

Peter Woit[10] in his 10 years inside as a String Theorist writes "String theory is not a theory; it's a hope for a theory." Although Woit does not offer another approach, he nonetheless makes a compelling case as to the wrong path that String Theory took sucking most of the physics theories into it like a black hole for several decades.

---

[10]   Peter Woit, *Not Even Wrong*,

How can we populate the metric tensor with dimensions without defining subspace or having what is known as compactification?

This is answered below in more depth. But briefly, this is manifesting physically by more fractal-like sets of 4D spacetimes, making a composite spacetime of co-existing 4-manifolds, where matter is the "other" spacetime. We can populate the metric tensor with the additional spacetimes as required by the physical observations, groups of 4 dimensions are defined side by side, 4 more columns, and rows in the tensor. These 4 groups are comprised of 3 more columns dimensions for the transformation space and 1 more column for the various time dimensions, plus other dimensions from which curved space needs to transform space and spacetime curvature between the two spacetimes. Hence we can model 13 dimensions in the metric tensor, similar to the number in TOE. However, more spacetimes may exist as well from the different 4D spacetime of the nucleons and sub-nucleons. The fractal-like sets of spacetime metric tensor are presented in more detail in a later chapter.

What is Electromagnetism, EM?

As alluded to above, EM is fundamentally the phenomena of the transformation of spacetime curvature, between the HGSMT and EGSFT spacetimes, inside and outside of the atom respectively under relativistic effects. This warps the immediately surrounding space in such are way as to make straight lines appear curved giving the appearance of a traveling two-pole "magnetic field", or an alternating inward space curvature push, magnetic field, with an alternating outward side curvature push, "electric field". EM is manifested as a transverse traveling wave, which has 3 oscillating spatial dimensions, just as they exist in the EGSFT, inside of matter. As the curvature of an atom must transform to HGSMT, and eventually relax into flat space with monotonic time away from matter, the spatial oscillations from circulating time must map to a monotonically increasing time, creating a traveling spacetime 3D spherical wave train which manifests as EM, traveling to infinity in a straight line as a pulse or train of Twisted Klein bottle morphed and rotating space balls. Since frequency-time circulates at a constant rate C inside EGSFT, the atom, the propagation speed of EM waves matches that temporal rate wrt atom space, and typically called the speed of light C, but is the speed of the traveling EM at rate C in HGSMT.

    Permeability and permittivity are the two constants used in calibrating the strengths of electric and magnetic fields. The square root of the reciprocal of their product strangely yields the speed of light. The speed of light c that does not directly depend on a measurement of the propagation of electromagnetic waves can be calculated using vacuum permittivity ε_0 and vacuum permeability μ_0 established by Maxwell's theory: $c^2 = 1/(\varepsilon_0 \mu_0)$. Oddly enough this is the constant in the spherical wave equation for spacetime curvature changes, giving the proportionality between rate of changes in space as affected by rates of changes in time and vice versa.

It is known that electrical, magnetic, and chemical forces are on the order of 10^{40} times stronger than gravity forces. Gravity as we know it is an artifact of the curvature of spacetime at the earth's surface. This same static curvature causing gravity from our sun was measured at 1.75 arc seconds of deviation from straight. Interestingly enough, the difference in curvature from the physical curvature of the starlight around the sun to that of a typical atom is of that same order of magnitude, 10^{40}. Since the spatial dimensions in the atom have corresponding frequency-time, an EM F-sphere train released from the atom, has sinusoidal changing X, Y, Z dimensions, which in propagation, comply with the spacetime monotonic time outside the atom, manifest as oscillations and rotations bound for infinity traveling ray like in a straight path. The physical reality is such because the EM train is a series of lobes or ellipsoids whose surfaces are normal at the propagation axis, producing the highest curvature between perpendicularly aligned lobes, and hence compel propagation of EM trains in straight lines along the lobe train rotation axis. More about light and photons are below.

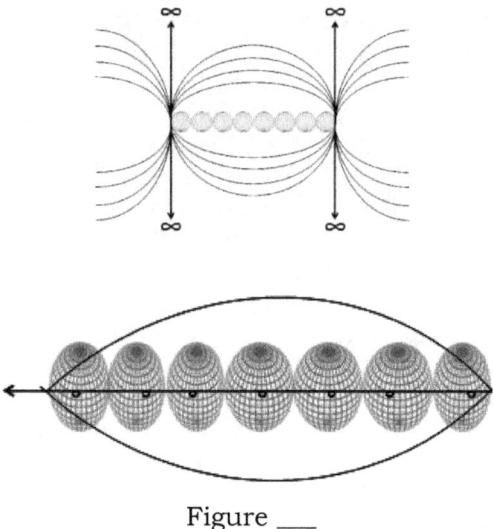

Figure ___

Higher relativistic speeds bring more contraction and hence more spatial bending around the moving spacetime sphere train.

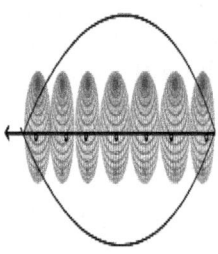

These then on a macro scale turn into a field with "bent" straight lines surrounding the string of atoms or "islands" of aligned clusters between two endpoints to distort or stretch a spacetime what is known as a magnetic field. In this fashion, the straight lines of space are warped by the contraction of space between the domain north and south poles into a familiar magnetic field configuration below.

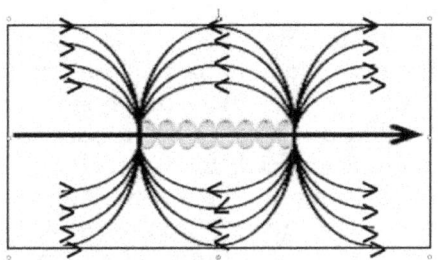

How did the time dimension transform from one 4D spacetime to another 4D spacetime, in the original formation of matter?

As shown above, with the mathematical transformation of a straight line from one geometry to another, Time was inverted in a Mobius transformation or geometric inversion to create another kind of spacetime, where time became a reciprocal time or frequency-time. This is shown below.

The time inversion created an "inside" and "outside" of a 3-sphere in a timesheet that contains itself. In accordance with non-Euclidean geometry, the sphere boundary is a one-sided surface. We will see later that this applies nicely with the Chronature atom model, as the inside and outside are made of a one-sided surface.

As the outside spatial dimension becomes the inside spatial dimension in the morphed Klein bottle, the time dimension transforms from monotonic to frequency-time. This physically represents the "prior art" to our 4D F-Sphere, and nature's way of transformation from one 4D to another 4D spacetime. We speculate that the big bang, the spatial dimensions travel faster than or pre-existed the time dimension, hence the time dimension curled up, in vortexes created by the expansion shear between these time and space dimensions. We will talk about spatial expansion later but the speed of time rate wrt space was C and needed to reach infinity ASAP. Curling up time finds infinity, and curled time is endless, with no beginning and no end. Hence as the universe is expanding, matter forms eddies of spacetime where time could reach infinity instantly, without being constrained by its rate of travel at C. This scaling fractal style of spacetime also solves the singularity problem of a black hole, which is in fact at 4D spacetime 3-sphere with frequency-time, just much much larger.

To talk about abstract concepts like the straightness or uniformity of time, we first turn to mathematics. This is mathematically completely analogous to the preservation of sphere inversion to 3D structures. Needham[11] shows that we may generalize the inversion of circles to straight lines and vice versa to spheres and planes. Hence complex variable mathematics gives us the foundation for the preservation of spacetime

---

[11]   Tristan Needham, *"Visual Complex Analysis"*, pg133

curvature by providing preservation of circles into straight lines and spheres into 3D sphere trains.

The following figure is found in Needham's[12] treatment of inversion of circles, transforming circles b, inside of a circle space K on the complex plane to straight-line outside, and vice versa.

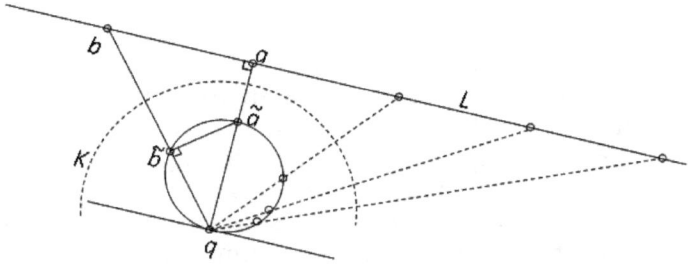

Figure 1.3

Complex Analysis would tell us if we had a circle within a complex circular plane, then through the process of Inversion we could mathematically flip it outside the circular plane and it would transform into a straight line or uniform time to circular time. Conversely, if we had a uniform straight line progressing uniformly and monotonically, we could theoretically, transform that uniform straight line to inside a circle of another space. So, mathematically we can transform time into cyclical repeating time or frequency-time. This frequency-domain time ultimately manifests physically as boundary harmonics. Since time and space are invisible but

---

[12] Tristan Needham, *Visual Complex Analysis*, pg 127

warpage effects are measurable, as we are still near enough to physical reality whereby the mathematical crutches help us to gain an understanding of the physical reality of this very abstract mathematical transform.

What is the mathematical model for an atom?

The basic physical model of an atom is that it is a 3-sphere in 4D manifold Riemannian space; the atom's boundary is the 3-spherical surface that follows the equations of the vibrating sphere. The atom is the spacetime within a spherical object. That object is the boundary between two spacetimes, the inside of the atom circulating time and drifting nucleons at the 3-sphere physical center. It is this spacetime curvature boundary between the two spacetimes, undulating or vibrating with the frequency-time inside and monotonically increasing time outside the two differently spacetimes whose curvatures converge at the atom boundary. The physical atom model is then the sphere boundary with frequency-time inside.

Why do atoms and matter appear "solid"?

The effect of cyclical time gives the EGSFT many different properties. For example, taking an object and ignoring Lorentz contraction, from outside, transitioning into the atom positive curvature space boundary will acquire a new position in the EGSFT atom space within which it finds itself. This new position in the atom's geometry changes the curvature from the outside negative curvature space as it enters another spacetime. Thus as it enters the atom's spatial volume, a new spatial position and the new time are automatically assigned to its new position. The new time changes the observed velocity of the entering object, in speed and or direction because velocity is measured as the change in position divided by the change in time. Furthermore, changes in velocity, delta V/delta T, change in velocity with a change in time, on the entering object are observed as acceleration. The new velocity and time are perceived by us as a force that the atom exerts on the entering object, $d(mv)/dt$. Hence the atom has a "solid" like property and with sufficient change in time, can be observed to bounce directly back into the direction of the original trajectory as the object enters, changes time, and leaves the atom boundary after having been spun around in time. But the backscatter force is due only to the new perceived velocity and perceived time change which changes the object's initial trajectory. Another way to understand this is that the entering object must comply with a spinning or oscillating space inside the atom volume, which is spacious with a cyclic moving time. The merging of new and old spacetime positions changes the trajectory of the object and it appears to have been deflected from something "solid", hence giving the atom a solid appearance, where it is only our perception from a changed trajectory, scattering of the encroaching entity.

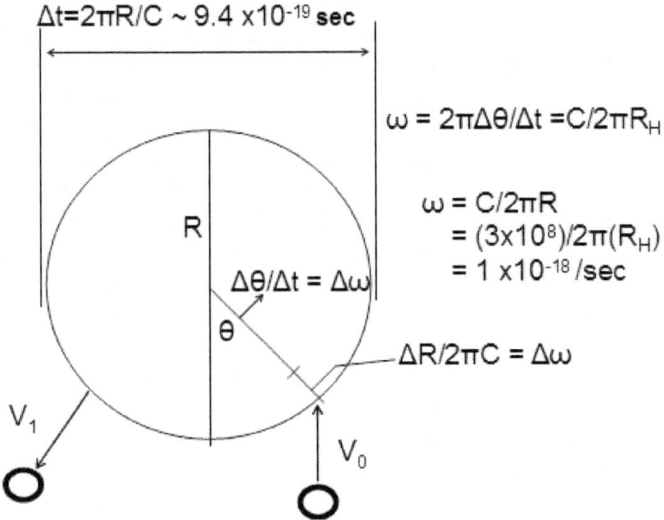

Atomic repulsion at spacetime topologically deformable atomic boundaries occurs mostly through impulse-momentum forces. Impulse **J** is defined as the integral of a force, F, from curvature acceleration over the time interval or time difference between the two spacetimes, Δt, for which the force acts. A small force applied for a long time produces the same change in momentum, the same impulse, as a larger force applied briefly.

$$J = F_{average} (t_2 - t_1)$$

The impulse is the integral of the resultant force (F), spacetime curvature acceleration, with respect to change in time:

$$J = \int F \, dt$$

Impulse- momentum, $\Delta\mathbf{p},$ is the change in linear momentum from time t_1 to t_2.

This is the classical definition of impulse force but is used here to calculate the repulsive force of physically contacting atoms at atomic boundaries.

Atoms by their negative spacetime curvatures will always be attractive, as in the classical definition of gravitational attraction. Upon contact, they will be repelled by an impulse-momentum force J. The time interval involved can be the time that the atom average boundary displacement from just pre-boundary touching to the elastic atom surface boundary springing back against atomic spacetime encroachment from another spacetime entity. Time censure is the more technical Minkowski spacetime discussion words.

Conservation of spacetime dictates that spacetime curvature is preserved and the topologically elastic atomic shapes must deform to preserve the integrity of the boundaries unless the boundary is breached in some manner, mostly under light speed conditions whereby time dilation effects cause the encroaching particle to approach zero time. This interaction between atoms is the main phenomenon giving rise to the thermodynamic character of gases and liquid states.

What is the physical nature of matter as taught today?

Tegmark[13] recites the physicist's typical understanding of the most fundamental building block of matter, a particle, as "it's an element of an irreducible representation of the symmetry group of the Lagrangian." This is probably way too abstract for normal comprehension. Classically matter is built out of a hierarchy of particles, Standard Model, without a clear understanding of the physical reality of such particles or the hierarchy of matter. Most theories of matter get deeper into the mathematics of one model or another, further departing from a physical understanding. The physical models have long gone by the waist side with using a logical fallacy argument of QMT that it cannot be known by any reality that we can perceive and that a particles unknowable position and velocity make it impossible to understand physical reality, and that it can only be understood to a probability with the actual physical mechanism unknowable to any certainty. Furthermore, we are pressured by the physics authorities that we must have faith to believe that this particle can be in multiple places at the same time, which is a physical violation of the Pauling Principle but still a reasonable scientific process.

The presently accepted Standard Model classifies atom components as "sub-atomic particles" named and properties measured, and this comprises atoms. Molecules are various amalgamations of smaller particles, protons neutrons electrons. Smaller particles are further amalgamations of even smaller particles, bosons, leptons, and quarks which are further down charmed, strange, up-down, colored. Analogous to an old poem about fleas by Augustus De Morgan:

---

[13] Max Tegmark, *Our Mathematical Universe*, pg 165

"Great Fleas have little fleas
    upon their backs to bite 'em.
        And little fleas have lesser fleas,
            And so on ad infinitum..."

And likewise with matter, that it is comprised of yet smaller unexplainable pieces and particles. But why is this so and what are these ever lesser fleas? String Theory teaches that matter is a mode of vibrating strings or membranes in 11 or 26 dimensions. And they vibrate at certain and discrete frequencies. Since they are orders of magnitude smaller than even the smallest Standard Model particle, on the order 10^{-26} cm, we have virtually no chance in heaven or hell of knowing what those strings are physically or why they vibrate and why at certain frequencies. Furthermore, how those strings interact to form larger strings and eventually form matter is unexplainable in any real or physical sense. But mathematics becomes curiouser and curiouser until it is without form and void of understanding. Perhaps the strings get tangled up and then form a bond. But why don't the loops get tangled up and then how do they get untangled if they do? This must be unsatisfying to others besides just me. Somewhere in space and time, that argument fails credulity and we are asked to have faith in incorrect models. Nothing against faith and hope but science is founded upon physical structures and functions which can be modeled in mathematics, and frequently tested for consistency and repeatability. It seems that modeling physical phenomena got turned around, and mathematics became a method to fit physical hypothesis with mathematics and was lost in what Richard Feynman[14] called "hocus pocus". The scientific humble honesty of Feynman is still unprecedented, when he spoke of the most complicated of theoretical models he admitted that he "did not understand how mother nature did it," and they could only show us how to "count the beans" as it were, to make predictive calculations.

Models of the real physical nature of matter are necessary if we are to ring out every feature of the mathematical model to advantage for a scientific understanding of reality. The Chronature model is foundational on the theoretically defined mathematical consistent field structures, with given names and entity labels made long ago. But these various named fields are based on what we see and measure physically of reality. Our true understanding of how nature builds from mathematics starts in GR, where matter causes spacetime to warp, gravity force fields magically appear. For years we could only make crude calculations for relatively slow speeds via the Newton model. The would-be acceleration fields out of the thin air and unseeable. But moving forward with the actual model of a force of gravity manifesting automatically with spacetime warp, the various fields can all be related to physically measurable spacetime curvature. Hence with a specific configuration of spacetime curvature, any known physical phenomena can explain any field, by the mathematical constructs representing spacetime curvature in a human abstraction or thought alone. But it's not chicken-or-egg, the spacetime curvature came first, not the other way around. Mathematics

---

[14]   Richard Feynman, "QED – *A Strange Theory of Light and Matter*"

does not create anything; everything is created by spacetime curvature and that includes matter. Furthermore, in the absence of spacetime curvature, we have nothing but flat spacetime, avoid and without matter as we know it. So for the duration of this book, we will study the topological construction and manipulation of spacetime curvature to explain all of the physical forces and fields that we experience.

Are there any physical atomic models that can aid in understanding?

The current physical models use subatomic particles comprising an atom with protons and neutrons in the center and one or more orbiting electrons whose exact position is not known since the electron lives in a "probability cloud", a pre-designated lobe volume which the "orbiting" electrons favor in space. This was an "improvement" over the Rutherford "Plum Pudding" model. QM was later used to explain the uncertainty principle application as a non-physical "probability" extension.

Yet the quantum cloud contribution was considered an "improvement" over the Bohr model which had the electron orbiting in some prescribed orbit each with "constant angular momentum." We all know both of these are incorrect, not physical and unreal "adjustments". From all their current models, physicists are at a loss to explain why the electron is not pulled into the center or nucleus by the proton, a much bigger stronger charge opposite and attracting charge. By divine decree, the kinetic energy of the electron was pronounced a stable constant, to keep the electron balanced in this physically unjustified atomic model orbit. In all the debate it is easy to forget that we have no good physical model of the atom that works or at least one that describes what we see and measure to any precision. Like a worn-out theme to fall back on for physicists, Quantum Theory is based on a model of the hydrogen atom, but it is not based on anything physical or real, using instead a probabilistic approach which asks believers to take on faith that "angular momentum" miraculously of an ethereal particle is conserved, for the electron orbital model to correctly predict the electron shell structure "interpreted" in an unreal sense. Moreover, this approach builds in complexity quickly to provide physical parameters for larger atoms with errors that need correcting corollaries to explain the discrepancies from measurement. So the answer is, not really. Although some help us in making some calculations and some predictions.

What do we know about matter for certain?

Current theories claim that atoms are neither particle nor wave, but have a particle and wave dual nature, with what we have come to take on faith from older models as a nucleus and electron(s) orbiting the nucleus in some manner yet to be discovered. But it's kind of like the three blind men describing the elephant. The current models of the atom argue for electron "clouds", orbits and probabilities surrounding a much more massive "nucleus" center. We know matter has the wave packet duality, but these are a characteristic of matter as well as energy. There are many characteristics of matter that we do know, but a good model governing all of them is not yet available, and hence the impetus for the Theory of Everything. TOE has made it so there is currently no way to prove or disprove TOE with current physical measurement. Moreover, it is an intellectually unsatisfying theory of matter since it explains nothing real and is based on more unexplained entities like strings, **brans**, and loops. What are they made of and how can they be used to create a subatomic guitar is unknown?

    Many believe that this current modern theory of matter is not only severely flawed but that the newer theories of strings, symmetry, supersymmetry, superstrings, GUT, and TOE are all tangents chasing after higher dimension space tail to explain what can't be explained with these types models. TOE is the one large, super theory, to "rule them all", with an unknown quality called matter in it. And the reason that all those GUTs are on the same path is that they add a column or two to the Metric Tensor, and rely on mathematic for some way to produce all of the forces that we measure. And so we should not be so quickly convinced that just because one can add a column and row to the Metric Tensor and turn the magic mathematical crank, that we can have a valid model to aid in our physical understanding of nature. Even bad models, like a broken clock, can sometimes give the correct result.

    Tegmark[15] points out correctly, fundamental particles "at the deepest level appear to be purely mathematical objects." And this would bring us full circle to the Chronature model, that everything that is anything is just another form of curved spacetime,

---

[15] Max Tegmark, *Our Mathematical Universe*, pg 166,

the spacetime of Einstein and Minkowski. That matter is a composite hierarchy of curved spacetimes, spacetimes in sets of one temporal dimension and 3 spatial dimensions, said dimensions from a set of different geometries, inversions, curvature transformations, and in general mathematical rules exclusively.

What does General Relativity say about Matter?

In discovering his theory on gravity, Einstein postulated that spacetime must be warped, having negative curvature in hyperbolic non-Euclidian geometry space in the proximity of matter. That was mostly for amalgamated massive clumps of matter such as celestial bodies. The corollary is that the more massive body of matter produces the larger negative curvature of space and spacetime producing a larger gravitational force in telling them how to move. Hence the more pronounced the spacetime curvature in the proximity of a body the stronger the gravitational field generated by said body. But all this can be modeled equivalently with a point mass, all of the mass can be represented by a point in the model for the calculations.

Thus with sufficient spacetime curvature, we would have sufficient mass trapping even light passing its event horizon, creating a singularity at the center, also where the modeled point mass is located, of a Black Hole. We know from GR that the curved geometry of spacetime warp is what causes attractive gravity. However, the Black Hole singularity introduces problematic infinities, since singularities have infinite depth and black holes have finite dimension. Matter "warps" spacetime, creating gravity forces by virtue of geometry change. Thus after Einstein, we know that the accepted theory of space, that it was indeed Euclidean, was wrong in the proximity of matter. Hence our current treatment of all physical laws, which all fundamentally rest on flat time-space Euclidean calculations and models, are only muddy partial theories at best. It was, in accordance with General Relativity discovery, a Riemannian Hyperbolic non-Euclidean geometry of spacetime warp that was the cause of gravity.

Starting where GR leaves off, we posit, that matter is also nothing more than curved spacetime, but organized under a different non-Euclidean geometry, Elliptical non-Euclidean geometry. The two known non-Euclidean geometries, Elliptical and Hyperbolic, must transition and co-exist from inside to outside matter and beyond. Hence, our time-space model in vast reaches of galactic space which rests on GR is good as proven, but completely askew on the nanoscale near and inside matter.

Why is matter "solid" and how can an atom just be empty spacetime, and still repel an object?

The Chronature model holds that matter is comprised of atoms with a temporal difference inside from outside, existing substantially from the disparate spacetime curvature. Hyperbolic Geometry Monotonic Time, HGMT, spacetime presently modeled outside the atomic spacetime, interacting with Elliptical Geometry Frequency-time, EGFT, spacetime on the inside will cause acceleration to an impinging object due to the difference in spacetime curvature over the time interval that the impinger is within the inside boundary. Within the atom boundary of space, a positive curvature and spacetime with circulating time will act to accelerate a traveling through entity during the travel time. Just as an object is attracted under gravity without an external force, solely through the progression of time in a hyperbolic geometry curvature spacetime, circulating time inside a positive curvature spacetime will induce an object inside to be compelled to the center of the space volume. Thus nucleons, themselves smaller 3-spheres of another disparate spacetime, are pushed toward the center of an atom, through what is called the Strong Force, repelling gravity from the boundary inside. Objects entering the positive curvature space of an atom's spacetime will change trajectories because they will occupy another kind of space curvature and acquire different times. A change in velocity traveled over a change in time will be the acceleration force applied. If all space coordinates are given a temporal component, as a circular time will affect, then any position inside an atom will have x+wct, y+wct, and z+wct, as time cycles at speed c. The constancy of time inside the atom, c, is also why EM travels at C, light speed, constant in a specific medium. In spherical coordinates, we would have $r + wct$, $\theta + wct$ and $\emptyset + wct$, where $0 <= r <= t$ and c is time speed.

$\quad$ Thus an object with changing spatial position, Δx, from a reference of translation in macro or outside space, will have a velocity measured from HGSMT. Where that translation leads one object to intersect with another object, eg. atoms, the curvature of the spacetime boundaries will have and intersecting EGSFT 4D volumes. Inside that intersection, the topological boundary will determine which time coordinates of space are applicable. The position will remain relatively the same for both objects, to each geometry and hence spacetime curvature differential. Since the boundary of each disparate spacetime will remain nearly the same in its own space, sny translating or "moving" object having a velocity with a change in time will appear to experience acceleration, $\Delta v / \Delta t$. Thus acceleration on an object will be accelerating from a change in direction. That acceleration "force", from the object taking on a different spacetime, will impart a new trajectory. This may well be additive to the interacting sum of the object's spin vector which will appear for all intents and purposes and be perceived by an observer, as scattering from something "solid" or a collision with something "solid". That is scientists believe that an atom, the fundamental unit of matter, has **a** substantive repulsing collision form. As there is a collision and scattering occurs. The collision and scatter is nothing more than the interaction of at least 3 disparate spacetimes, the two atomic entities in colliding in the surrounding macro spacetime. Since the interaction will depend on the size of each atom and the shape of the curvature and the curvature between, as well as the initial positions and velocities in macro space in which the atoms hang, the scattering matrix will differ for every single collision. Most collisions at low velocities push particles apart but still not out of "gravitational" attraction range, turning collisions into stable but vibrating-rotating bond mechanisms; with increased "temperature" ie translation, vibration, rotation, these collisions increase and with decreasing "temperatures", these interaction decrease. That is the very definition of "temperature".

What is the physical mechanism for time transition from monotonically increasing, outside the atom, to cyclical, inside the atom?

Felix Klein showed how through two Mobius transitions, combined in bottle format, the outside surface can become the inside surface. Time has a very similar twist but in a 4D Klein bottle configuration. As we know, there is a difference between the three spatial dimensions and the temporal dimension. But spatial are moving relative to each other. Therefore there is a "stress-tension" manifesting a viscosity between time and space dimensions in adjustment, even at time c. Time must reach infinity in synchronicity with space, as at the Big Bang, time transitions into a sphere vortex of sorts, or a spherical spacetime, a 3-sphere and in this way instantly reaching infinity, morphing into a localized spherical shaped spacetime sheet. Thus space and time "viscosity" transforms spacetime into spacetime spheroid vortices or matter as we call it.

On an oscillating 4D, spacetime surface, where inside the time oscillates and the spacetime boundary has a dynamic curvature, a stationary object moving through space will be morphed into an oscillating spatial volume, from monotonic to frequency-time change imposing a discrete change on the trajectory of the moving object. On a unit circle, any circulating time component will appear as a circulating spatial component from a non circulating monotonic increasing time spacetime, and the spatial component is subject to the elliptical geometry space with rotation about each component due to the circulating time dimension inside the Positive

Curvature Space (PCS) atomic boundary. Hence the spatial components inside the atom appear in our "monotonic time" to oscillate in synchronicity with the circulating time inside the atom. Likewise the spatial dimensions inside subatomic particles appears probabilistic, that is "particles" live inside probabilistic space volumes, are only the affect of their spherical volume oscillates continuously, giving them the appearance, characteristics and our perception of a wave. But its curvature boundary makes it appear as a particle since circulating time will change an encroaching objects trajectory by assuming a difference in time changing velocity, giving the appearance that the object was deflected.

A 3D Klein bottle is shown below but the atom is a 4D spherical spacetime Klein bottle. The negative curvature or flat space and monotonic time spacetime is transformed into a 4D spherical Klein bottle of positive curvature space (PCS) with circulating time (CT). Frequency or cyclic time is perceived as providing the attribute to "solid particles", as any outside object entering this atomic spacetime will instantly be subject to a change in coordinate space, giving it the net affect of an new trajectory, effectively scattering or repelling. Note that the 4D Klein bottle emanates from the atomic center, and has symmetry (with sufficient topological morphing). These aspects of the atomic structure, providing the many properties and attributes of this model, because the atomic boundary is dynamic and interactions with outside objects would be time averaged from the outside-in and time described realm inside-out.

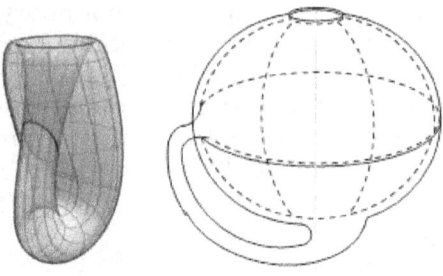

The rules of Elliptical or Spherical Non-Euclidean Geometry provide some of the model answers. First, a surface inside the 3-Sphere, atom, Elliptical non-Euclidean Geometry Space (EGS) sheet has only one side. Physically in a 2-sphere, the spacetime sheet will perform at least two Mobius transformations to flip the outside surface into the inside surface. See the Klein bottle above for two Mobius flips giving a 3-sphere. However, taking the time dimension from the outside and mapping it into the inside of the 3-sphere by inverting the time, creates a 3-sphere with a frequency-time, or inverted time. This is a physical model of a purely geometrical structure in 4D that can provide spherical harmonics where the openings are sufficiently morphed tightly.

The volume inside an EGS sphere will likewise be vastly different from a sphere composed in Euclidean geometry. The area and volume in Euclidean space increase with 2 and 3 power, respectively. Whereas inside the EG sphere, the power is much higher. This indicates that space inside the atom will be highly "concentrated" or "compact". Transformation of this spacetime to outside the atom spacetime will therefore have a spatially increasing effect, pushing all surrounding space to allow accommodation for more new space.

We currently perceive and measure this delta energy in fission or fusion of matter as "mass defect." as the space immediately surrounding the fissioned atoms changes spatial positions at time speed C, changing any objects in the altered space, outside the atoms, different spatial coordinates in time speed at time speed.

What is the Boundary between Spacetimes, EGSFT and HGSMT?

Since the negative curvature space (NCS), Hyperbolic geometry space, co-exists with EGSFT spacetime within atoms, NCS must warp around the positive curvature space (PCS) and hence we view this as "matter warps space". This is shown graphically below. That which we can visibly observe and that currently postulated to comprise space, must transition to PCS, elliptical geometry, that which we cannot see or measure without great difficulty. Using the following: a spherical model for a boundary, a revolving spherical atom with a spin axis, a discrete step transition between the negative curvature hyperbolic, and a positive curvature; elliptical non-Euclidean geometry and associated non-Euclidean rules, we have an approximate model for predicting many properties of an atom. The center of the sphere will have a time gate, by virtue of its vibrating sphere harmonic modeshapes, for the transition from one geometry space to another, i.e. one spacetime to the other spacetime. This goes part and parcel with our measurements and perceptions of the thing we call an atom. I posit that since at the subatomic level, space is compressed into 3D elliptical volumes and cycle time in those 3D elliptical volumes, since they operate in PCS, we perceive them to have density and weight and hence "mass", as the higher spatial density and cyclical time appears to have more "inertial resistance" than the surrounding adjacent less curved hyperbolic non-Euclidean space with time sheets to infinity. This "inertia" is a direct result for spacetime drag due to interaction with local surrounding spacetime curvature.

General Relativity or as also known, the theory of Gravity, tells us that spacetime can bend and is warped around matter. Spacetime warp occurs without kinks, following rules of Topology, and can be modeled by various topologies with properties, and in conformance with mathematical rules from hyperbolic non-Euclidean geometry. But since time, to be shown, somehow becomes transformed from monotonic to curved, inside matter space, the space-space and spacetime curvature of the atom conforms to a different set of mathematical rules, providing the physical forces and attributes measured. Scientists measure and record these as physical properties of matter. But they are wholly predicable, with a true physical model of the atom with appropriate geometries.

The figure below shows some modeshapes for a circle harmonics. These are equally applicable to spheres and are shown here to illustrate the point that spacetime is dynamic, that spacetime boundaries and hence spacetime curvature is dynamic. Thus the properties if atoms and all atomic "particles" are determined by spacetime curvature boundary and immediately adjacent spacetime curvature within proximity of other spacetime boundaries comprising matter. As we leave the spacetime boundary the adjacent spacetime curvature reduces at and hence the forces are attenuated at roughly:

$$\sum_{W=1}^{N} K_W / r_W^2$$

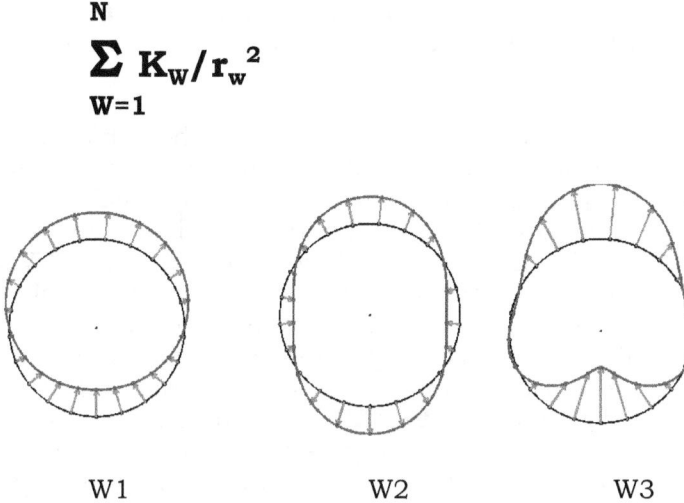

W1　　　　　　W2　　　　　　W3

QM with Bohr model and angular momentum constraints provides measurement validating predictions from a hydrogen atom, the major one being the electron shell structure. This result communicated to the physics world that they were on the correct track, with a good model. But I proffer that result to be serendipitous, because the equation derived from Schrodinger using constant angular momentum becomes identical to the equations for a vibrating sphere equivalent to the Chronature physical modelV3-sphere membrane between two different and separate spacetimes. The spherical wave curvature becomes the boundary of the atom, a membrane between two different types of spacetime. Hence the atom boundary is a well defined spacetime curvature with spatial harmonics as well as temporal harmonics, standing waves and potentially traveling waves. These can only transform at discrete normal independent frequencies, solutions to the physical model spherical wave equations.

The questions then turn to how thick is the atom's boundary, what is its density and stiffness of the boundary and what are the topological deformation properties of this membrane boundary dividing the two spacetimes we call the shell? What is the effective topological membrane made out of and how can pure spacetime curvature make such a discrete change in spacetime even possible? Part of the short easy answer is that dimensions are part of a model and that model represents real physical boundaries. An analogous phenomenon can be found in Boundary Layer Theory. Engineers studying fluid flow, adjacent to a solid structure, found that there was another world in the boundary between a fluid and a structure. Inside this "world" was a completely different flow regime and molecular phenomena which required different modeling to understand and use. By analogy the atomic boundary is a very small Boundary Layer Theory of Chronature, where spacetime geometries meet in the real physical world. These have a spacetime shear aspect to them which gives us the macro properties of the atom boundary.

Standing shocks are another analogous real world phenomena wherein the states of matter, gas or fluids, undergo discrete changes across a discrete change boundary.

Another way to approach and understanding of the boundary is from what is known about Black Hole and their event horizons. Many interesting things have been deduced to happen at the event horizon since the proper time, frequency-time inside the atom boundary, and real time, what we see and measure outside the boundary, will be very differnent. Moreover, the radius of courvature of the Black Hole is the inverse Swartzchild radius squared. So the atomic boundary ramifications are the bigger the atom, element atomic number, the smaller the outside boundary curvature and hence smaller the attractive force from the boundary curvature.

What is the current position physics take on spacetime curvature and the 4 natural forces of gravity, EM, Weak Force and Strong Force?

Since GR established that the spacetime warp causes and explains gravity, Einstein's GR explained and superseded Newton's mass action at a distance theory of gravity, although Newton's equations for the calculation of gravity between masses is still used and very powerful for most real and for large scale mass physical problems. The remaining three known forces are still not explained by a pure spacetime geometry model like GR for the gravitational force. In fact, the four forces are classified by their relative "strengths" and are reported by Feynman[16]. Of interest to us here, as we will make a cursory check of that later, is the relative strength ratio measured between gravity and other "forces" measured by

---

[16] Richard P. Feynman, *Six Easy Pieces*, pg 44

"coupling" to be of the order of 10^{-40}. Feynman quickly and most importantly adds with humility, "we seem to be groping toward an understanding of the world of sub-atomic particles, be we really do not know how far we have yet to go in this task."

    More will be covered on how to calculate the force of gravity from spacetime warp, but we have some measurements which can help us to get at least a quick relative feel for the forces exerted by space warp from warpage around our sun, and a hydrogen atoms spacetime curvature/ geometrical properties.

1 – Geometries of Spacetimes

Physical spacetime rests on the postulates and rules of Geometry, but which ones?

The nature of space is defined by spacetime geometry which rest on rules, one of which is that straight lines must always be straight in the space and the time that the line traverses as per the fifth postulate. In flat spacetime a straight line is the straight line, the shortest distance between two points, the line that we learned in school ala Pythagoras. **Euclidean geometry.** That is a line travels straight, in accordance to the parallel postulate, from infinity to infinity with zero thickness. There, a parallel line cannot pass through any points on that line without crossing the line. This is in stark distinction with hyperbolic geometry, where a space or time is curved so that an infinite number of straight lines can go through the same point and still be parallel. Moreover the forgotten and little known Riemannian manifold or Elliptical geometry, straight lines are geodesics, the shortest distance between two points on elliptical non-Euclidean surface. In this space all straight lines on a surface, an infinite number of straight parallel lines. intersect at the two poles.

It is precisely these combinations of straight lines in the different geometries traversed through matter, that produce warped spacetime; Straight lines can be additive based on the geometries adjacent to each other, extending the straight line in one spacetime or composite of spacetimes wrt to other straight line traversing in their own but different geometrically governed spacetime.

The continuity of spacetime geometry demands that lines comply with a continuum, be continuous and monotonic, not kinked. As shown above, straight lines in non-Euclidean Hyperbolic geometry must transform into non-Euclidean Elliptic geometry great circles or geodesics. A corralary is that travel or transitions along a geodesic, a straight line in curved spacetime, will have special state change properties or measurements much like travel in straight line flat space.

In general, our perception of space can have devastating affects on reality. Riemann analogized this to when the crinkles in spacetime exert physical forces on a traveler.

What is the affect of geometry on spacetime?

Geometry is how we navigate through spacetime. The Pythagorean Theorem fails in all non-Euclidean geometries. Anything but flat space creates a new "metric", ie your direction of straight will bend.

The physical manifestation of right angles leads us to the mathematical notion of curvature, the differencing parameter of geometries. We can only understand spacetime effects by modeling and calculating spacetime curvature. The Gaussian formulation for curvature identifies the geometrical bending at a point and is defined by the product of a surface's principle curvatures, which occur at perpendicular axis. The scalar curvature κ at a point on a surface is defined as

$$\kappa = + (1/R_m * 1/R_n)$$

where the R s are the perpendicular to each other axis principle radii tangent to a point on the surface. A surface of a flat plane has curvature zero, $1/\infty$, but in general, the radius is inversely proportional to curvature, the greater the magnitude of curvature κ, the smaller the radius tangent to the surface at that point. The sign tells us qualitatively what the surface is like in the immediate neighborhood of the point. Examples in the figure following show that where curvature, K, is negative, we have a saddle point, positive where the surface bends outward in all directions as from inside of a sphere surface, and zero where flat.

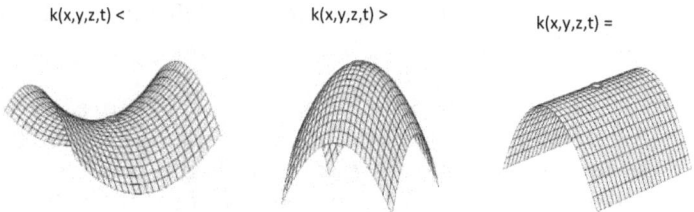

Beltrami[17] gave us that "Euclidean, non-Euclidean spherical, and non-Euclidean hyperbolic geometry can all be interpreted as the geometry of constant

---

[17] Tristan Needham, "Visual Complex Analysis", pg 274-275

vanishing (zero), positive or negative curvature respectively."

How does this help us to understand spacetime affects and effects? We can't see space and we can't see time. But we can map an imaginary grid onto the curvature and visualize the spacetime curvature where there is a discrete change in curvature. As the boundary of an atom forms a dividing edge of spacetime, we will look for the mathematical solution for the atom boundary shape to calculate spacetime curvature. The shape of and curvature on one side of the atom boundary membrane will provide the force of attraction to that surface, just like gravity. Why? Because spacetime curvature is acceleration, change in an object's velocity due to a change in direction in moving time alone. Acceleration generates a "force", or what was known as "mass action at a distance." Where Newton's formulation to calculate the fore could not explain the underlying principles, GR does just that and with more accuracy.

In genera on a 3D surface, curvature is inversely proportional to the tangent radius squared. so larger radii of curvature will have a smaller attracting surface or gravitational force. See the diagram below which shows that the smaller sharper surface will have greater curvature and hence stronger attraction force than the larger surface. Reason being, $1/R_1^2$ is smaller than $1/R_2^2$. Thus eccentricity on objects of spacetime curvature will exhibit charge, that is some surface boundary regions will be more attractive than other regeons based on their curvature in spacetime. Thus charge is purely a spacetime geometrical property. Thus one atom's particular modeshape surface will be "electropositive" to the other or vise versa, "electronegative," depending on the resonance, and or synchronicity between the entity boundary spacetime surface curvatures.

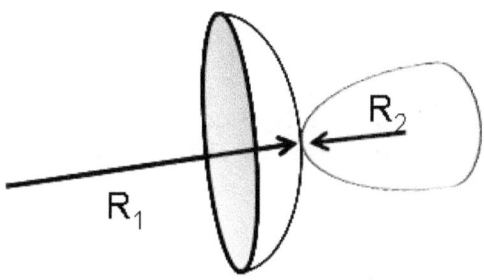

Once a spacetime surface is modeled in computers, giving us the curvature over all points on a surface, we will no longer be dependent on experimentation for answers. Moreover, the affect at some distance will require the summation of the spacetime volume subtending that affected point. The one big advantage of this method is that we are not required to undertake the enormous task of determining the Stress Energy tensor and calculating the Einstein Field Equation to provide answers in a specific case for a specific point on an atomic spacetime surface. But also because we can choose the geometric configuration for certain phenomena, we can greatly simplify for computational models, the complexity of the tensor calculus required for interacting 4D spacetimees.

We can take advantage of choosing where we stand when we make the calculations, much like Eratosthenes of ancient Greece in his calculation in

Alexandria and Sudan in calculating the circumference of the Earth. Eratosthenes measured a shadow cast at midday at the specific place on a specific time on exactly the same day and hour a year apart. Where we "stand" in space and time, our reference frame, are both vital and necessary for geometrical spacetime measurement.

How weak is "gravity" compared to the other of nature's known forces of EM, Weak Force and Strong Force?

Since GR established that the spacetime warp causes gravity, Einstein's GR superseded Newton's theory of gravity. Newton's equations are still useful for calculating gravity forces most problems. The significance of GR is what it did to our thinking is what was the dramatic change. Moreover the remaining three known forces in the Universe which have formulas for calculation are yet physically not explained by a uniform elegant model like GR for the force of gravity. In fact, the four forces are modeled and classified by their relative "strengths" and are reported roughly by Feynman[18]. Of interest to us here, as we will compare results from a spacetime curvature perspective, that there the relative strength between gravity and other forces like chemical and electrical forces is to be of the order of 10^{-40}. Physicists comparing the strength of the classical gravity force to the strength of ordinary electrical forces find that gravity is 10^{-40} times smaller than electrical forces, thereby excluding it as the same force. Feynman quickly and most importantly adds

---

[18]   *Six Easy Pieces*, Richard P. Feynman, pg 44

with humility, "we seem to be groping toward an understanding of the world of sub-atomic particles, be we really do not know how far we have yet to go in this task." From results out of CERN it would appear that physicists are still "groping" as models for the "ballpark" still give us little clues as to physical structure of matter.

Calculating the strength of the electrical forces as if they were produced by spacetime warpage, using a simple first order curvature calculation, and comparing that to the curvature of gravity forces generally applied to gravity, like our sun, we can obtain the relative strength of gravity to electrical forces, using spacetime curvature. The bending of light around our sun was measured at 1.75 arc seconds, curvature around the Sun. We can obtain the scalar curvature by using the sun's radius and the measured spatial starlight bend, the sun's radius is 7×10^{10} cm and 1.75 arc seconds. We know that the radius of Hydrogen is approximately $.5 \times 10^{-8}$ cm, assuming for the Chronature model is valid and the atom boundary is precisely spacetime curvature, the relative strength ratio between traditional gravity and electrical energy using curvature at a given on a spherical object surface point is:

curvature κ_H/ curvature κ_{sun}

~ C/R_H^2 / C/R_{sun}^2

~ $(7 \times 10^{10} \text{ cm}/\cos(1.75 \text{ arcsec}))^2/(.5 \times 10^{-8} \text{ cm})^2$

~ 10^{39}

where C is a constant of light speed.

Using very simple average space-space curvatures to calculate relative force or energy strengths, we get approximately the same result as the traditional relative strength measurements provide. This could be mere coincidence but also goes further in establishing a general validity to a unified field theory using curved spacetimes.

Why can we use the radius of the sun in our spacetime curvature comparison?

Einstein applied the Sun's radius in calculating the curvature of starlight passing, in his famous answer to Eddington's confirming measurements of bent starlight during an eclipse. He got lucky because he chose to use the spacetime component of curvature to calculate the space-space component of curvature. So the real answer is a little more involved because spacetime curvature has compoents. Spacetime curvature has two components or mathematically speaking, the sum of the partials, spacetime curvature partial wrt time and spacetime curvatue partial wrt space. In the case of the Sun's gravity, the largest component of curvature comes from spacetime partial wrt time, 28g's, as time has to travel the diameter of the sun of 1.4×10^{11} cm at 3×10^{10} cm/sec for roughly 14/3 seconds, and not spacetime partial wrt space proving a grand total of 1.75 arcsec of bend. This is shown below. And so we used the Sun's radius for the curvature comparison as it provides the largest spacetime component, acting immediately to accelerate an encroaching entity rather than 5 seconds to bend direction by 1.75 arcsecs.

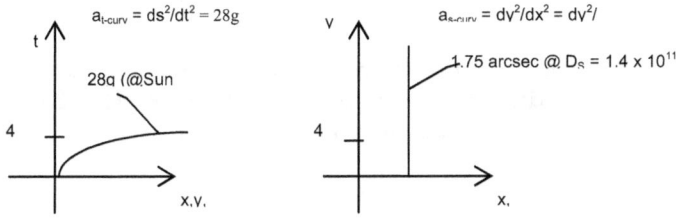

It is no coincidence that the gravity curve you see is a parabola. This curve is a straight line in non-Euclidean hyperbolic geometry and its second derivative, rate of change of rate of change geormetrically speaking, turns out to be a measure or value of acceleration, manifested as a physical field. Thus accelration is related to curvature. Just another clue that gravity is not much more than geometry change.

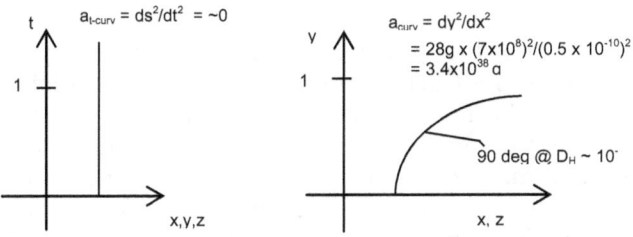

Add harmonics for the temporal-spatial modeshape contribution to spacetime atom boundary metric and quantization is born, but above sum of curvatures shown represent the total from the partials in graphical form from the Chronature model geometry of a vibrating sphere boundary of spherical spacetime.

Why do we have to complicate and obscure everything with tensor calculus of EFE, when we can still use Newtons gravity law?

In spacetime, generally time is a moving dimension or at least "sytem time" is. Another no small thing is coordinate systems, which one do we use?

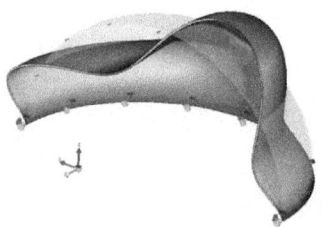

Coordinate systems can vary, and everyone's favorite is the Cartesian Coordiate system which is made for flat space and flat spacetime, ie where there is no matter. But nature seems to enjoy making all kinds of matter, so there follows energy and changing coordinate systems. This is the reason for using tensors and tensor equations. They provide solutions for any geometry and take advantages of things called "invarients" that don't change based on coordinate system changes.

But also we can calculate the acceleration of gravity by geometry alone. And notice at no time do we need mass or the distance between masses required by Newton. We simpley graph the time an object will "fall" against the falling distance in the proximity of matter. On a space vs time axes, the straight line will form a curve, hence curved spacetime, a bending of an axis in the Euclian space. That graph is a special geometrical curve known as a parabola or paraboloid, conic section, whose 2nd derivative will provide a number, the magnitude of gravity as acceleration.

How did time physically get inside the subatomic level elliptical geometry spacetime?

In spacetime as we know it, time is a moving dimension, sometimes graphed as ct. Time is moving relative to space at rate c, speed of light c, one second of distance per second. However time dimension in the atom spacetime is looping or frequency or inverse time.

In an elliptic geometry plane, a straight line does not divide the plane into two distinct regions nor does a surface into volumes. Elliptic geometry physical surface closed surface sheet has only one side. That is it must be comprised from the elliptic geometry which defines a plane as a one-sided surface[19]. Physically this will impart a 4D Klein bottle

---

[19] D.M.Y. Sommerville, *The Elements of Non-Euclidean Geometry*, pg 92

character on the subatomic boundary it's spacetime. Where there is a twist in the surface the curvature, creating both a positive and a negative curvature on the same side of a surface, simultaneously creating a "north" and a "south", attractive and repulsive curvature respectively.

Turning to the macro time dimension momentarily, the generally accepted conception of time is embodied in the Steven Hawkings articulation that time is like an arrow that goes in one direction forward. But the Minkowski diagram would argue that time has a negative aspect. Moreover there is the Feynman observation that matter moving forward in time is equivalent to anti-matter moving back in time. But this forward and backward is still uni-dimensional. We posit that time within a subatomic positive curvature elliptical spacetime complies with a different spacetime geometry for both its spatial and temporal dimensions. Speculating on subabtomic particles inside of a atom's 3D sphere in elliptical spatial geometry of spacetime of a neutron, there spacetime is governed by elliptical geometry and time is cyclical, the bi-directional and infinite characteristics of time are preserved in elliptical geometry in combination with in a negative frequency ($-1/Time$). Moreover, average time remains stationary relative to the embedding composite spacetime, since time dimension circulates continuously inside the subatomic "particle". This on the average makes time in a neutron remain zero our time, that the quark, a neutron's first harmonic, does not age. Our speculation is that frequency-time in a subatomic particle is reversed or has higher sub frequencies or higher harmonics, creating a totally different spacetime from that inside the atom spacetime.

However, as matter is accelerated to the speed of light, its inertial reference from time will also slow to zero, see the Minkowski diagram above. As this happens, the matter may physically be undergoing Lorentz deforming "stresses" on its boundaries which deform the curvature past its diffeomorphic, isomorphic, and or homeomorphic topological boundary physical limits resulting in a boundary breach. EM may result as EM is the transition of spacetime curvature from one kind of spacetime to the other, converting frequency-time to light speed monotonic time, with much higher EM frequencies.

A traveling entity must give up its local time from the elliptical to the hyperbolic spacetime configuration of space. This spacetime difference and transition from elliptical geometry positive curvature space to hyperbolic geometry negative curvature space are what we observe as gamma and even higher frequency photons, as they start as spacetime curled up in a ball and unravel, taking the reflective and rotating space curvature from within the atom outside transitioning to another spacetime geometry. It is the rotation and reflection character in elliptical space frequency-time that give rise to the oscillations in a hyperbolic spacetime, because the time dimension must "travel or move" to reach infinity, from being localized to a very small region of space with a closed tail for the benefit of reaching infinity. This transition or relaxation from one spacetime geometry to the other, when encompassing the entire particle, must transition between the alternate geometries, a transformation of spacetime, which is evidenced as "energy". That transformation is what we observe as a perturbation of surrounding regions of ellipsoids causing translational, rotational, and vibrating atomic boundary sphere undulations.

So to answer the question, how did time turn on a dime, accomplished a transformation from one geometry to another at the speed of time, c? Whatever the physical trigger is, it is/was all done in accordance with the rules of mathematical and geometrical Inversion, using existing or forming curvature wrt proximity to each other either in space or in time or both. The moving aspect of time wrt space in all cases provides the curvature needed initially as a movement of curvature via time provides the velocity and adjacent geometries in proximity provide the change in direction, together with giving a change in velocity, in turn, acceleration and thereby forces. Like a river or stream near a shore, the fluid where "slowed" by the land, forms pockets of eddies and vortices, circulating volumes of the same river fluid. Thus spacetime vortices were formed and are still so doing at the edges of the Universe.

2 – A Least Two or More Spacetimes

What are the components of Spacetime?

The short answer is at least 4 dimensions comprise a spacetime, that is three spatial and one temporal. But that is not even the interesting property of spacetime, curvature is because the spacetime curvature is directly related to acceleration and hence forces. The dynamics of spacetime curvature and bending dimensions are what make spacetime introduce forces. Flat spacetime is probably the vast majority in the universe but somewhat boring since not much happens there. So this leads us to what is curvature, which's so great?

By the Einstein Equivalence Principle, we know that there is no difference between an acceleration field and a gravity field. And since the temporal dimension is moving, acceleration fields can be set up just by virtue of curved spacetime. The concept of spacetime curvature lies at the heart of GR and our understanding of the force of gravity, what it is and how it works. Likewise, the calculation of curvature and geometry resides at the heart of Chronature, as the models of matter and energy are all nothing more and nothing less than spacetime curvature and transitions between spacetimes. Later we will introduce a spherical spacetime whereby the time dimension is moving cyclically, causing an acceleration field inside of the sphere we currently call the Strong Force.

 Intuitively, the curvature is the amount by which a geometric object deviates from being flat or straight in the case of a "line", in the context of the 5th or Parallel Line Postulate. Bear in mind that a change in the Euclidean 5th Postulate creates an entirely different geometry. And since the very geometries of space and time are difficult enough to understand beyond the abstract, we mentally resist their physical ramifications. However, our treatment of curvature here will not be mathematically rigorous but just enough to paint the physical reality from the unseen and abstractions of differential geometry in dimensions of space and time.

 Einstein's discovery as to how gravity works physically, lead the world's interest in the concept of curvature. It is our responsibility to figure out which geometry we need for a particular aspect of nature and choose our models accordingly. Chose correctly and the models are relatively simple and elegant, choose arbitrarily and models become inaccurate, require more variables, and behave strangely because they are not compliant with measured results.

 Einstein's GR simple calculation of starlight passing near the Sun did not receive the benefit of his Field Equations, as he was able to by simple model obtain numbers that would be used to verify GR. This is shown in the diagram below:

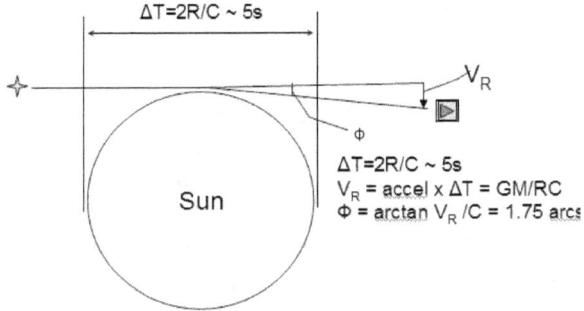

He calculated roughly an angle of deflection d:

$$d = 4GM_s/RC^2$$

where G is the gravitational constant, M is the solar mass, c is the velocity of light, and b is the minimum distance between the trajectory and the center of the Sun and the radius of the Sun is R_s, we have:

$$d = 1.75'' \, R_s/b$$

These were confirmed by the famous Eddington's solar eclipse measurements. However, these results were disputed by some and claimed that the fundamental physical principles involved in that deflection of light were violated. That is it was found that such a deflection is not compatible with the principle of mass-energy. Other theories were proposed and studied, but none seemed any better as the results came out about the same. We include this here to show that Einstein was fully capable of making a simple calculation in a simple physical model as to the value of spacetime curvature and found that a photon passing by the sun would be warped 1.75 arc seconds.

As for the Sun in the experiment, curvature used is loosely confined to and for objects embedded in space and or time, usually Euclidian and monotonic respectively, at first but in a general non-Euclidean space or time co-existing.

In a measurable, numerical, and physical sense, curvature relates to the radius of curvature of circles that touch an object or surface, and "intrinsic" curvature, which is defined at each point in spacetime, eventually a Riemannian manifold. We deal with curvature primarily with a spacetimes radius of curvature at specific points for ease of understanding.

The canonical example of extrinsic curvature is that of a circle or a hypothetical circle tangent to a curve. This hypothetical radius tangent circle has a defined curvature equal to the reciprocal of its radius. Smaller circles bend more quickly from their tangent and hence have larger curvature, 1/radius. So generally for any curve or surface, the curvature is defined as the curvature of its osculating circle(s) reciprocal radius at any point on the curve or surface.

The numerical value of the curvature for any point on a curve is a scalar quantity, but one may also define a curvature vector that takes into account the direction of the bend as well as its steepness or deviation from the tangent at a particular point on the curve. Mathematics was developed to keep track of the curvature of more complex objects such as surfaces in more n-dimensional spaces. These are more efficiently managed by more complex mathematical tools and objects such as the general Riemann curvature tensor. Because for the cases of sphere harmonics, we will know curvatures of surfaces of interest to numerical precision, we will be able to make good predictions by inspection of where we would need the curvature calculations on a surface.

We further throw a monkey wrench into the rigorous mathematical wheels of curvature when we bring a time dimension instead of another spatial dimension to the manifold. The math of curvature will barely cover the basics needed to more closely model and predict the physical nature of matter and energy from spacetimes curvature alone, much less a pseudo-Riemannian manifold to include the time dimension. From a practical standpoint, this will go back to the basic definitions of force and acceleration.

To add a little graphics to our treatment of surfaces and surface curvature, let C be a plane curve shown below. The curvature of C at a point is a measure of how sensitive its tangent line is to move the point to other nearby points.

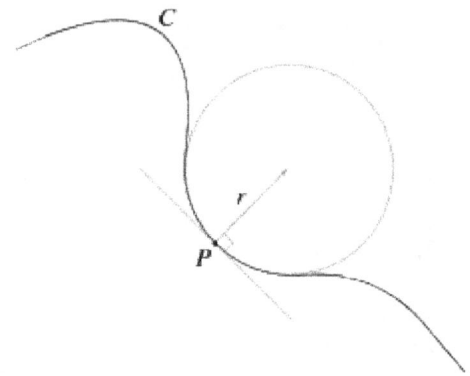

Accordingly, the curvature of a straight line is numerically zero. Numerically the curvature of a circle of radius R would be large if R is small and small if R is large. Thus the curvature of a circle is defined to be the reciprocal of the radius:

$$\kappa = \frac{1}{R}.$$

Given any curve C and a point P tangent to it, there is a unique circle or line which most closely approximates the curve near P, the osculating at P. The curvature of C at P is then defined to be the curvature of that circle or line. The curvature at a point is defined as the reciprocal of its radius of curvature.

The sphere is an especially unique and simple object for defining curvature and is recognized as having convex curvature. The sphere, as we will see, is a major component of the structure matter as we know it.

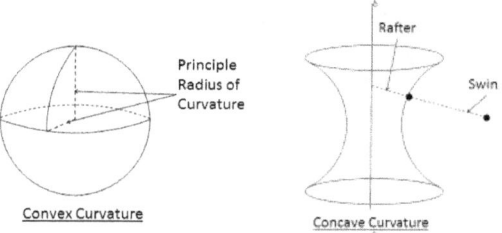

On the other hand, the curvature can also be negative or concave, for example, the hyperboloid[20]

---

[20] *A Journey into Gravity and Spacetime*, John Archibald Wheeler, pg 139

shown above has negative curvature. As we shall see an atomic entity can have both positive and negative curvature, forming a dipole.

So pushing onto curvature in the time dimension, our intuition would tell us that time always marches forward and with a uniform tick tick tick. With this hardwired understanding, it's barely comprehensible that time is bent or non-uniform. So seeing a straight line represented with curvature on a Minkowski axis would appear intuitively wrong since that would mean time ticks are not uniform. But one of the greatest scientific discoveries of all time calls for just that time dimension curvature.

So what if the straight-line curved or a curved straight line had a curvature different from another straight line on the same plane or surface? Keep in mind that we are talking about empty space but we can legally model any region with straight lines of any geometry, so we need to fully describe its fundamental spacetime character is through its geometry. Having done that it would be said that one straight line has "acceleration" from another straight line because it has curvature, and not due to any inertial external force, we could say they were still on parallel paths.

What does that even mean? If a curve remains little changed in a direction, the unit tangent vector changes very little and the curvature is small; where the curve undergoes a tight turn, the curvature increases dramatically. Bear in mind here we are discussing any curved surface, which will have a principle axis or 2 straight lines perpendicular to each other, but in our model, the curves are straight lines or flat surfaces in the spacetime geometry that they are in. This is enforced by the 5th Postulate in any given geometry.

Going back to physical reality for a practical application, atom boundary surface is a sum of spherical harmonic surfaces from a series of special functions defined on the surface of a sphere, see the above chapter on the electron structure, and provide solutions of differential equations. As Fourier series is a series of functions used to represent functions on a circle, spherical harmonics are a series of functions that are used to represent functions defined on the surface of a sphere. So there you have it, the relationship between the spacetime curvature of an atom and its mathematical underpinnings using harmonics ala Fourier. This makes the vector geometry and algebra simpler because we know apriori the numerical nature of the curvature enclosing any atom without the use of the Einstein Field Equations and Stress-Energy matrix to find numerical answers in describing all of the forces involved in the vicinity of a said atom.

How does spacetime curvature cause acceleration hence a physical force we call Gravity?

This is after all the question that was answered by GR, what is gravity, and how does nothing or nothingness manifest a physical force?

Acceleration is simply stated merely as an instantaneous change in velocity. But not so simple, velocity is a vector that is velocity vector has magnitude and direction. So acceleration is a change in velocity, which means that there is a change in magnitude or speed or a change in direction of velocity, or both changes in speed and direction of velocity.

A continuous curve C can be describe as a function of spatial x, y, z, and temporal dimensions

position r(t) $\quad$ = r(t)i + r(t)j = r(t)k

velocity v $\quad$ = r'(t) = f'(t)i + f'(t)j + f'(t)k

acceleration a = r''(t) $\quad$ = r''(t)i + r''(t)j + r''(t)k

a = **v**'(t) = d**v**/d**t** = v' + dθ/dt

$\quad$ where **v** is speed and **θ** is direction. The reader will note here that direction can change by virtue of external, centripetal, force on the particle or by entering a space with a change in geometry, ie. in going from flat Euclidean to a Hyperbolic non-Euclidean curved spacetime geometry.

Therefore the velocity of a moving particle on C in its principle components tangent and normal

$\quad\quad$ **T** = r'(t)/|r'(t)| , **N** = d**T**/dt/|d**T**/dt|

Speed v = ds/dt = |**V**(t)| , Velocity **V**(t) = v**T**

Acceleration = a(t) = v d**T**/dt + dv/dt **T**

The acceleration of the point P on a curve C can be designated as the change in the tangent vector d**T**/ds over the path length s, or the rate of change of the rate of change of a point's position on a differentiable surface.

Curvature κ = |d**T**/dt|/ds/dt =>

$\quad\quad$ dT/dt = κ v**N**

$\quad\quad$ a(t) = κ v²**N** + dv/dt **T**

Geometrically, the curvature κ measures how fast the unit tangent vector to the curve rotates in the principle axis directions. Alternatively

$$\mathbf{a}(t) = a_N + a_T$$
$$a_N = \kappa\, v^2 \mathbf{N}\ ,\quad a_T = dv/dt\ \mathbf{T}$$

This indicates that a change in velocity can occur in the time-space plane, xyz-t, or the space-space plane, xyz-xyz. Although we can get a change in speed and or direction from curvature in the time-space plane, we can only get a curvature from a change in direction in the space-space plain. But total acceleration will be the sum of all the components of change in velocity in the three distinct spatial dimensional planes. These accelerations can manifest from curvature alone.

Moreover, curvature of spacetime can come in the form of temporal dimension geometry change for direction. Hence an acceleration force emerges merely from the passage of curved time, ie. from a straight line curve in the spacetime geometry. Such a straight line curve provides the curvature from which acceleration emerges first in time followed by motion in space.

If spacetime curvature creates gravity, why can't we calculate the acceleration of gravity by just using spacetime curvature?

The short answer is we can and we're going to do just that. But exact formulas for spacetime (ST) curvature are not easy to find as they are invisible, and in most cases must be measured or mapped by some means, sometimes very expensive means. When we do know the exact ST curvature, they come in two flavors or components. Where the contribution to ST curvature is from the dimension of time, we can calculate the force of acceleration without Newton, without knowing the mass of the earth or object to be accelerated, ie. without the mass of the attracted object, without the Gravitational Constant and the radius between objects. We can calculate the acceleration of gravity from spacetime geometry alone.

First, we calculate the acceleration component from ST curvature wrt to time. We need a graph or map, a simple measurement of a falling object subject to the field would do it. The graph below is the result of a few measurements plotted on the space vs. time axis of a falling object at sea level. Just to highlight as stated above, gravitationally accelerated object's mass, a Gravitational constant, the mass of the earth, and the distance between the object and the earth are irrelative. These are requisites to the Newton formulation. For a quick example, we will use measurements, in units of feet, of an object falling for 1, 2 and 3 seconds rounded respectively to roughly 16, 64, and 144 feet.

For simplicity, our measurements will plot space vs. time as shown. This plotted curve turns out to be a parabola, a well-known conic section in geometry. So part of the solution is that instead of using the heavy machinery tensor calculus of the EFE, we take advantage of our knowledge of the geometry of the spacetime field itself from its geodesics. For example, the geometry of a parabola. The equation of a parabola is well known and plotted this way starting at the origin and aligned with the space axis is

$$t^2 = 4\,p\,x$$

t = time,
x = distance (space)
p = parabola focus length

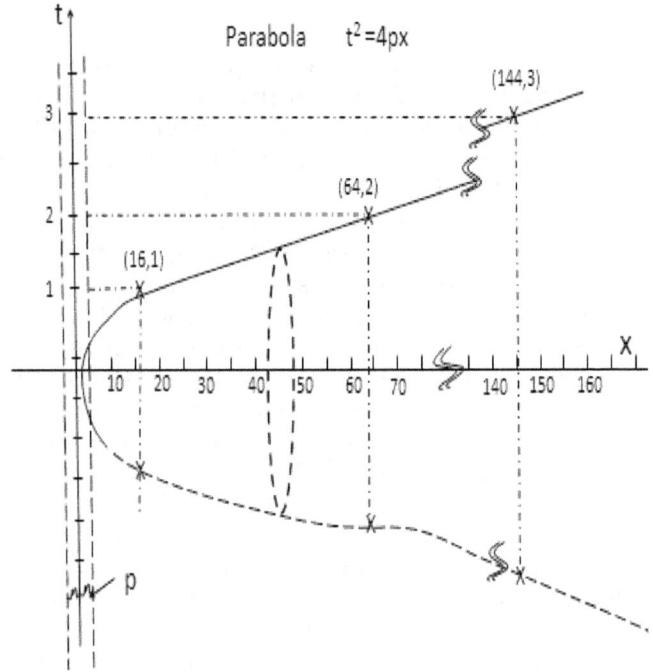

A parabola's focus length p, swing, rafter, directrix properties, and other conic section geometry in spacetime are well explained by Wheeler[21] and only simplified here, but the presentation in Wheeler of the geometry and calculations of spacetime curvature for various geometry objects encountered in GR is well worth the read.

We have good reason to know that the spacetime geometry of the earth's gravity is a parabola. Let's start plotting the spacetime geometry of the earth's gravitational field with a few measurements.

Putting a curve point into the equation, (1) ^{2} = 4p (16), yields p = 1/64

The radius of curvature for our parabola is 2p at zero, the osculating radius directly from the focus for the parabola, as we seek only dependence wrt time. From above we know that the spacetime curvature K wrt time is the reciprocal of this radius, hence

$$K = 1/2p = \text{acceleration} = g = 1/2(1/64) = 32 \text{ fpss} \ (= 980 \text{ cm/sec}^2)$$

g = acceleration of gravity at earths sea level, zero
K = curvature = 1/2p

Putting this with the parabola variables, the acceleration of gravity is therefore roughly 32 feet/sec^2. Putting this together

$$t^2 = 4\,p\,x = 2(\,1/32\,)\,x$$

this can also be written for x as

$$x = 16\,t^2 = \tfrac{1}{2}\,K\,t^2 = \tfrac{1}{2}\,g\,t^2 \quad \ldots\ldots$$

Why does this work, that the graph of a spacetime position, a geodesic (straight line in the graph), yields acceleration?

For those with knowledge of vector calculus, the 2nd derivative of position in time yields acceleration. Thus the 2nd derivative of the spacetime parabola curve ($t^2 = 4px$) wrt time will also yield acceleration g.

$d^2x/dt^2 = 1/2p = g$ acceleration of gravity w/o knowing mass, G constant, or distance between them, d.

QED the acceleration of gravity from spacetime geometry alone, and at no time did my hands leave my wrists.

Why does this even work, despite the complexity of the EFE?

We have simplified the equations of motion to a very well-known mathematical formula, a paraboloid. Also, we have taken the largest contributor of the spacetime curvature, hence acceleration, Earth's acceleration field from spacetime curvature wrt time, and discarded the other contributor to spacetime curvature, that is the partial of ST curvature wrt space. As we shall demonstrate in the next calculation, for atomic spacetime dimensions, the bulk of the forces and fields will come from the ST 2nd partial derivative wrt the spatial dimensions, not the time dimension.

For those with a calculus background, the **geodesic equation** of motion is:

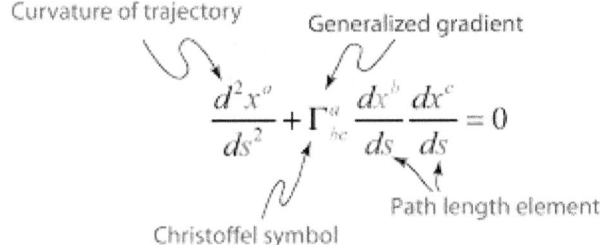

Geodesic Equation of Motion

$$\frac{d^2x^a}{ds^2} + \Gamma^a_{bc}\frac{dx^b}{ds}\frac{dx^c}{ds} = 0$$

(Curvature of trajectory — $\frac{d^2x^a}{ds^2}$; Generalized gradient; Christoffel symbol — Γ^a_{bc}; Path length element — ds)

Wheres is a scalar parameter of motion (e.g. the proper time), and Γ is the Christoffel symbol. The quantity on the left-hand side of this equation is the proper acceleration, the acceleration felt by the particle in a more or less uniform acceleration field.

If spacetime curvature creates gravity, why do scientists deem that the acceleration of gravity is different from the other 3 forces and not caused by spacetime curvature?

The short answer is because the acceleration of gravity is at least 10^{-33} times smaller than the strength of the other 3 forces and for this reason, spacetime curvature was deemed too weak to be the same force/field as gravity. For our next acceleration field, that atom, our calculations, and the result are due to the ST curvature wrt space, not time, for atom-sized dimensions.

The equation below is analogous to Newton's laws of motion which likewise provide formulae for the proper acceleration of a particle in a geodesic or straight line $\underline{S}$ for both space and the time dimension contributions. For us, this is:

$$\frac{d^2 S}{d \tau^2} = \frac{d^2 S}{d t^2} - c^2 \left(\frac{d^2 S}{dx^2} + \frac{d^2 S}{dy^2} + \frac{d^2 S}{dz^2} \right)$$

^	^	^
ST Curvature wrt proper time	ST curvature wrt Coordinate Time	ST curvature wrt space

A quick inspection shows that the curvature and acceleration are the same components by component. Where there is but a single atom in Minkowski spacetime, an atom's spacetime curvature produces surrounding acceleration field by bending the surrounding spacetime, a field which is produced due to the geometric distortion in the surrounding Minkowski spacetime. For talking purposes, a test particle is used, it is an entity whose own field is negligible and does not affect an atom's surface acceleration field, so it does not change the atom's surface acceleration field. But an atom in another atom's field, molecules and matter, change each other's fields, and this is currently known as "bonds". There seem to be as many "bonds" as elements and compounds are comprising these, due to a large number of dynamic spacetime interactions between this space and time distorting entities.

We have shown a typical ST curvature calculation for large masses, where the time's dimension contribution is the dominant term as in astronomical bodies. At the atomic level, space contributes to the spacetime curvature is by far the dominant contribution to spacetime curvature. Hence:

$$\frac{d^2 S}{dt^2} = 0$$

Second, we calculate the acceleration from ST curvature wrt to space. Here, we know that the acceleration component wrt time is very small. And we don't need to measure the curvature since an atom can be represented by a vibrating sphere and we know the formulas for that. And let's just take that for this calculation the sphere is static. What would the contribution to the total spacetime curvature, surface acceleration, be from just the spatial dimension of that atom? Assume that the atom radius is constant a.

$$\frac{d^2 S}{d\tau^2} = c^2 \left(\frac{d^2 S}{dx^2} + \frac{d^2 S}{dy^2} + \frac{d^2 S}{dz^2} \right)$$

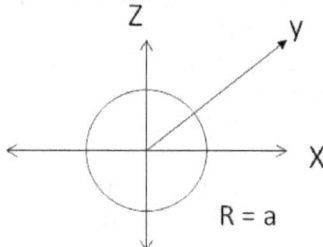

For one dimension, the 2nd derivative of the curve in ST will provide the osculating radius of the curve., needed for calculating the acceleration attraction force. For a sphere of radius a:

$$S = x^2 + y^2 + z^2 = a^2$$

at x = 0, we solve for the radius of curvature for the osculating sphere, 2-dimensional sheet, with the orthogonal x and y dimension radii of curvature multiplied

$$\frac{d^2 S}{d\tau^2} = c^2 \left(\frac{d^2 S}{dx^2} * \frac{d^2 S}{dy^2} \right)$$

$$d^2S/dx^2 * d^2S/dy^2 = 1/a * 1/a = 1/a^2$$

$$= (3x10^{10} \text{ cm/sec})^2 \times (2) (1/(0.528 \times 10^{-8} \text{ cm}))2$$

$$= 6 \times 10^{36} \text{ cm/sec}^2$$

This equals approximate acceleration at the Hydrogen atom surface due to ST curvature, wrt spatial dimensions

Please note, this result is entirely derived from the geometry of ST curvature, an atom embedded sphere, with the assumptions that the time dimension is essentially zero and the ST curve osculating sphere in the atom is for a sphere of radius 0.528×10^{-8} cm.

We can likewise calculate the acceleration for the surface of any element for which we know the geometry in space alone, given a set of harmonic mode shapes defining the atom spacetime for particular resonances or even average from a Fourier sum in time.

Comparing with the strength of the spacetime curvature comprising gravity g;

$$\underline{d^2 S} / dt^2 = 980 \text{ cm/sec}^2$$

Using the same coordinate units, the acceleration of gravity is very approximately 6×10^{33} times weaker and hence appears like another kind of force. Today these are called bond strength, electronegativity, electron affinity, Van der Walls forces, and other names. These are all just applications of spacetime curvature to a specific configuration of "subatomic particles"

For the math guys, how can we get more comfortable in the curvature calculations?

For a plane curve given parametrically in Cartesian as $\gamma(t) = (x(t), y(t))$, the curvature is [22]

$$\kappa = \frac{|x'y'' - y'x''|}{(x'^2 + y'^2)^{3/2}}$$

where primes refer to derivatives with respect to parameter t. The signed curvature k is

$$\kappa = \frac{x'y'' - y'x''}{(x'^2 + y'^2)^{3/2}}$$

The expression reflects the geometric meaning discussed above, that the curvature is influenced by the change amount of the tangent vector in the direction of the normal vector, as

$$\kappa = \frac{m}{(x'^2 + y'^2)^{3/2}}$$

$$m = \frac{x'y'' - y'x''}{(x'^2 + y'^2)^{1/2}}$$

---

[22] https://en.wikipedia.org/wiki/Curvature

The curvature of a graph

For the less general case of a plane curve given explicitly as y = f(x) and now using primes for derivatives with respect to coordinate x, the curvature is

$$\kappa = \frac{y''}{(1+y'^2)^{3/2}}$$

Consider a parabola $y = x^2$. We can parameterize the curve simply as $\gamma(t) = (t, t^2) = (x,y)$. If we use primes for derivatives with respect to parameter t, then

$$x' = 1,\ x'' = 0,\ y' = 2t,\ y'' = 2$$

Substituting and dropping unnecessary absolute values, get

$$\kappa = \frac{x'y'' - y'x''}{(x'^2 + y'^2)^{3/2}} = \frac{1 \times 2 - (2t)(0)}{(1 + 2t^2)^{3/2}} = \frac{2}{(1 + 4t^2)^{3/2}}$$

It would seem that curvature, here numerically acceleration due to change of geometry, would decrease with time.

Curvature: Partial Derivatives

Spacetime is a 4-dimensional continuum so obtaining the curvature at any point will require contribution from all the components of the dimensions as they change wrt each other. In our spacetime, we ignore the possibility that geometry can change, Euclidean space. But this is only an assumption that helps us make calculations quicker. Spacetime geometry changes and the change wrt each dimension must be accounted for. This is where General Relativity comes in. But on a simple level, we can think of the curvature at a point from the position of spatial dimensions changing wrt spatial dimensions with the additional curvature from spatial dimensions changing wrt time dimension. We call thise space-space and spacetime curvature respectively.

Curvature: Negative, Positive, or Zero?

The sign of the spacetime curvature was shown above for three cases, plus, minus, or zero. The sign of the curvature k indicates the direction in which the unit tangent vector rotates as a function of the parameter along the curve. If the unit tangent rotates counterclockwise, then $k > 0$. If it rotates clockwise, then $k < 0$. So, for example, the sign of the curvature of a graph is the same as the sign of the second derivative.

The signed curvature depends on the particular parameterization chosen for a curve. For example the unit circle can be parameterized by $(\cos(\theta),\sin(\theta))$ (counterclockwise, with $k > 0$), or by $(\cos(-\theta),\sin(-\theta))$ (clockwise, with $k < 0$). More precisely, the signed curvature depends only on the choice of the orientation of an immersed curve. Every immersed curve in the plane admits three possible orientations. As mentioned above, this is complicated a tad when the surface geometry has only one side, as in elliptical or spherical geometry, and whether that surface twists, thereby changing from an attractive curvature, lines accelerating away, to a repulsive curvature, lines accelerating towards.

Curves on surfaces of spacetime

But straight parallel lines on a surface only define the geometry that we are in and we need to know how this all turned into a real physical force or solid entity. A bunch of one-dimensional curves on a two-dimensional surface embedded in three dimensions creates curvature which spans an area and can create a volume and that is where we find the workings of physical nature. It is not the volume, it's the curvature of the entire volume at each point on the surface. Rigorous will require we identify a surface's unit-normal vector, **u** for all the different line curvatures. These are given the names of normal curvature, geodesic curvature, and geodesic torsion. Any non-singular curve on a smooth surface will have its tangent vector **T** lying in the tangent plane of the surface orthogonal to the normal vector. The normal curvature, k_n, is the curvature of the curve projected onto the plane containing the curve's tangent **T** and the surface normal **u**; the geodesic curvature, k_g, is the curvature of the curve projected onto the surface's tangent plane; and the geodesic torsion or relative torsion, τ_r, measures the rate of change of the surface normal around the curve's tangent.

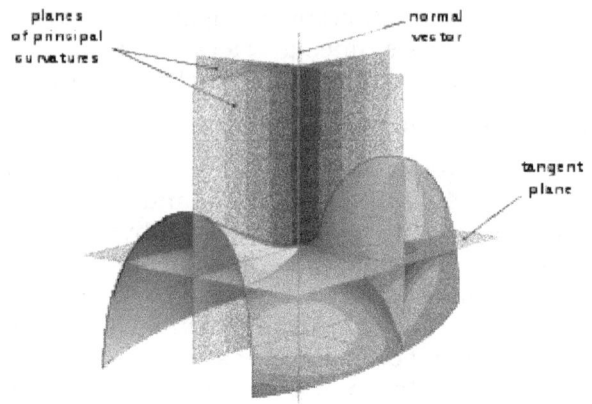

A saddle surface with normal planes in directions of principal curvatures is shown above. This can be a chunk of space between two atoms. All curves with the same tangent vector will have the same normal curvature, which is the same as the curvature of the curve obtained by intersecting the surface with the plane containing **T** and **u**. Taking all possible tangent vectors then the maximum and minimum values of the normal curvature at a point are called the principle curvatures, k_1 and k_2, and the directions of the corresponding tangent vectors are called **principal directions**. In later discussions, we will relate this to the two constants measured in nature, called permeability μ_0 and permittivity ε_0. Together the permittivity ε and permeability μ of a medium determine the phase velocity $v = c/n$ of electromagnetic radiation through that medium:

$$\varepsilon\mu = \frac{1}{v^2}.$$

Many tools have been developed in the extensive study of curvature of surfaces and are provided in the differential geometry of surfaces, dealing with "smooth" surfaces with various additional structures, the Riemannian metric. Surfaces have been extensively studied from various perspectives: *extrinsically*, relating to their embedding in Euclidean space and *intrinsically*, reflecting their properties determined solely by the distance within the surface as measured along curves on the surface. Surfaces will be important to us since they form the boundaries or enclosures separating disparate spacetimes, hence their embedding in Euclidean and non-Euclidean space become important in the quantification of physical properties, so how we choose to model the spacetime will determine how good our physical understanding andd predictions will be.

Curvature of space

Space of three or more dimensions can likewise be intrinsically curved. The curvature is intrinsic in the sense that it is a property defined at every point in the space, rather than a property defined with respect to a larger space that contains it. In Chronature models, a curved space may be embedded in the same numerically dimensional but different geometrically dimensional "ambient space." and so its curvature is defined intrinsically.

In accordance with the teachings of GR, once a time coordinate is defined, the three-dimensional space corresponding to a particular time is generally a curved Riemannian manifold. But since the time coordinate choice is largely arbitrary, it is the underlying spacetime curvature that is physically significant, that manifests gravitational forces. So that Chronature model physical reality our assumptions of the curvature of space must comply with the mathematical assumptions of local isotropic and homogeneous conditions, so that all points and directions of space are indistinguishable. Where in the case of the sphere or hypersphere, the "osculating circles" provide positive curvature corresponding to the inverse square radius of curvature.

An example of negatively curved space, just outside of an atom spacetime boundary, is hyperbolic geometry. A space or spacetime with zero curvature, away from any matter, is called flat or Euclidean geometry. Although the Minkowski space was originally graphed as representing flat space, we have taken some liberties in our modified Minkowski graphs, by representing the warpage through straight graphed lines that are flat, but longer, with the grid showing the rotation of the inertial reference frame rotation causing all the known forces to appear through the progression of time dimension alone at first followed by another warped spatial dimension.

Many topologies are possible for curved spacetime, and as their complexity increases quickly, the "metrics" are the unique identifiers for the magnitude and sign of the curvature of space that the models represent.

The curvature of time in space

Frequency-time is not an original Chronature idea. In mathematical physics, a closed timelike curve [23], CTC, is a world line in a Lorentzian manifold, of a material object in spacetime that is "closed", that is returning to its starting point in space and time. This possibility was discovered by Willian van Stockum and later confirmed by Kurt Gödel. Using this concept Gödel discovered a solution to the equations of GR providing for CTCs and is known as the Gödel metric.

    Van Stockum published a paper that contains one of the first exact solutions in GR which modeled the gravitational field produced by a configuration of rotating matter, the *van Stockum dust*. Van Stockum first understood the possibility of CTCs, one of the pillars of Chronature submanifold spacetime. Furthermore, submanifold spacetimes were embedded within spacetimes much as all atoms, separate and disparate spacetimes, are embedded in a macro spacetime and coalesce into what we see as matter in their different states and subatomics in association with matter.

    Some have noted that if every CTC in a given spacetime passed through an event horizon, a property named chronological censorship, then that spacetime with event horizons excised would still be causally well behaved and an observer may not be able to detect a causal violation. We theorize above that this translation from CTC submanifold spacetime to our macro spacetime occurs through a geometrical transition of inversion that we experience physically as EM. This is in compliance with the general primary conservation of spacetime rule and inversion dynamic referenced above.

---

[23] https://en.wikipedia.org/wiki/Closed_timelike_curve

In Minkowski space, a light cone represents any possible future path of an object given its current state, or every possible location given its current location. An object's possible future locations are limited by the speed that the object can move, which is at best the speed of light. For example, an object located at position p at time t_0 can only move to locations within $p + c(t_1 - t_0)$ by time t_1.

This is commonly represented on a graph with physical locations along the horizontal axis and time running vertically, with units of t for time and ct for space. Light cones in this representation appear as lines at 45 degrees centered on an event, as light travels at 1 to 1, ct per t. In this diagram, every possible future point of the event lies within the cone. Additionally, every space point has a future time, implying that an event may stay at any location in space indefinitely.

Separate events are considered to be *timelike* if they are separated across the time axis, or *spacelike* if they differ along the space axis. If the object were in free fall, it would travel up the t-axis; if it accelerates, it moves across the x axis, space axis, as well. The actual path an event takes in spacetime is bounded by its light cone.

In flat spacetime examples, the light cone is directed along the Minkowski space vertical axis in time. This corresponds to the physically represented case that an object cannot be in two places at once, or alternately that it cannot move instantly to another location. In these spacetimes, the worldlines of physical objects are, by definition, *timelike*.

However, this orientation is only true of "locally flat" spacetimes. In curved spacetimes the light cone will be "tilted" along the spacetime's geodesic, as shown in the figure directly above, the warping towards matter or in the vicinity of matter as shown, increases the metric for a straight line. In the physical world, an object in free fall under these conditions continues to move along its local t axis, but to an external observer, it would appear to the observer as if the object is in orbit.

In acute cases, spacetimes with suitably high-curvature metrics such as an ellipsoid would have, the light cone can be tilted beyond 45°. That means there are potential "future" positions, from the object's frame of reference, that are spacelike separated to observers in an external "rest frame. From this outside viewpoint, the object can move instantaneously through space. Under these circumstances, the object would have to move, since its present spatial position would not be in its future light cone. Additionally, with enough of a tilt, there are event positions that lie in the "past" as seen from the outside. With a suitable movement of what appears to it its space axis, the object appears to travel through time as seen externally.

A CTC can be created if a series of such light cones are set up to loop back on themselves, so it would be possible for an object to move around this loop and return to the same place and time that it started. An object in such an orbit would repeatedly return to the same point in spacetime if it stays in free fall. Returning to the original spacetime location would be only one possibility; the object's future light cone would include spacetime points both forwards and backward in time, and so it should be possible for the object to go back in time under these conditions. This was shown graphically in more detail above under the Feynman observation for a positron as being an electron going backward in time.

What is "energy" at the atomic level, including nuclear and EM?

To explain all that we see and measure requires interacting spacetime surface curvatures. So we begin with at least two different spacetimes.

From the standpoint of spacetime curvature, energy is defined here as the perturbation or the transition of curvature between spacetimes, the conversion of spacetime curvature from hyperbolic geometry space- monotonic time HGSMT to elliptic space-geometry-frequency-time EGSFT, or in reverse order. ie. Transitions of curvature from negative to positive curvature, positive to negative curvature or some combination with flat spacetime cause "energy" as we observe.

Translating an infinitesimal piece of spacetime surface curvature about a point inside the atomic spacetime adheres to the topological rules of mathematics for particular manifolds[24].

The mathematical group theory topographical motions and transitions allowed in elliptic geometry are rotations and reflections inside the elliptical spacetime, thus "energy", Thus the atom is a spherical boundary between NCS and PCS, combined with rotating space and cyclical time inside the elliptical geometry space. This is subject to and

---

[24] Needham, *Visual Complex Analysis*, pg 124

manifests as the vibration, rotation, and translations when NCS transitions to hyperbolic defined general macro spacetime. Electromagnetic radiation is the transition result of closed curved spacetime lines of a spacetime relaxing to straight infinitely long space lines. Thus, unimpeded by matter, EM will travel forever, extending out to conserve the curvature from which it originated. The classic model of a photon shows that the E field is sinusoidal with a B Field at 90-degree phase shift, both going from positive to negative as they are both rotating. Inside the atom, all three are axis which is looping in time at the frequency ω, which when Mobius transformed into "flat" space, they appear as we see them traveling forever at speed C. Looping time must change to straight space, from circulating to monotonically increasing time and this gives the space curvature fragment photon its speed as it transitions from a positive space curvature bound in the "atom" to macro space outside the atom, seen as a photon. Therefore a photon is a positive curvature space fragment in circulating time spacetime transitioning to macro space in monotonic increasing time spacetime. This photon, as we will see below, travels as a sphere train exhibiting both wave-like and particle-like characteristics until it interacts with another spacetime curvature producing a display of "energy", "quanta" or "packet."

What is physically going on with all these spacetimes and why don't we just use the different geometries to get the correct numbers?

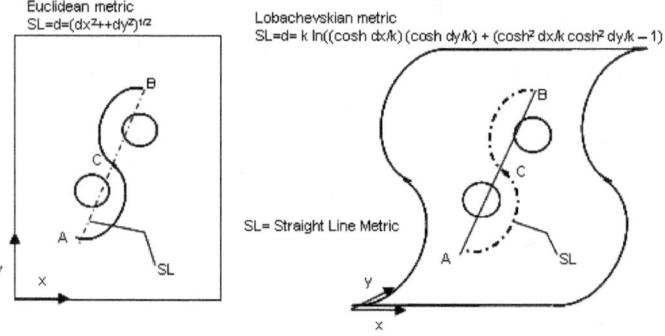

Fig 7.1a   FIG 7.1b

The figures above illustrate the importance of getting the differential geometry of a spacetime correct between atoms and groups of atoms. In a rectilinear or flat spacetime, FIG 7.1a, a straight line (SL) is shown as a dotted line and the metric or distance from A to B for two spatial dimensions is calculated using the Pythagorean formula we are all familiar with:

$$\text{Geodesic Length} = d = (dx^2+dy^2)^{1/2}$$

While measurements this small are difficult to make, they are being attempted never the less. The Casimir effect arises at nanometer proximities and there may be a good physical explanation from the Chronature model to explain the forces between very proximate very flat plates.

The distance in Hyperbolic spacetime geometry between A and B, shown in the dotted straight line in FIG 7.1b, would be calculated as:

$$GL^{25} = d = k \ln((\cosh dx/k)(\cosh dy/k) + (\cosh2 dx/k \cosh2 dy/k - 1)$$

Although the solid line in FIG. 7.1b appears shorter, it would have to be going through a 2D wormhole to connect A and B as it could not be in the plane or space shown because straight lines define the space that they travel in.

At very small distances the two metrics will converge to the same value at any local point, the large curvatures involved as shown at least in these figures, and as clearly shown by the straight-line lengths, SL, the distances would be substantially different between atoms. We bring this up here because any figure or diagram in the literature of atoms, lattice structures, and molecular models generally shows the distances within the atomic dimension neighborhood. Like the bending star beam bending by our Sun, spacetime curvature changed the position of the star from the apparent to the real, atomic and molecular structures are not modeled correctly, because of the different geometries in spacetime as we get closer to atoms. Now we can calculate the true positions. Atoms much more dynamic but organized in lattice structures makes for a potentially wrong calculated distance where the geometry is not accounted for.

At very large distances, as in the cosmos, GR shows us that curvature in spacetime as produced by very dense matter will bend a surrounding spacetime lattice in a predictable way, shown in the illustration below. Please note that although the lattice appears

---

[25] *Non-Euclidean Geometry*, Stefan Kulczycki, pg. 146

bent, the spacetime distances are all equal and straight-line geodesics between points.

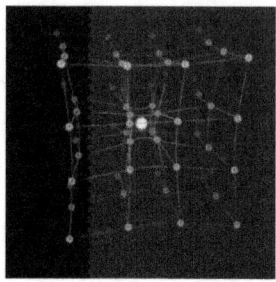

What is temperature?

The classic definition of temperature is kinetic energy of vibration, translation and rotation of atoms and molecules. Increases in any of the vibration, translation and rotation of atoms and molecules will generally increase temperature and decreases in these motions will cause decreases in temperature. Thus in all of these motions we need to determine precisely how the atomic boundary between atomic spacetimes reacts. Does it deform, stretch or does it breach and to what extent? Atoms and molecules in physical motion heat, give us clues and with some topological modeling we can eventually obtain good "engineering" numbers which will enable us to surgically move atoms to advantage without incurring the great random motion limiter Gibbs Free energy, statistically random thermodynamics penalty.

Also of interest, when it comes to understanding bonds, is what happens in the boundary between spacetimes when temperature is reduced, such as at near zero temperatures. What happens where atoms converge to the Bose-Einstein Condensate state of matter, were the interactions between 3-spheres of spacetime may combine or coalesce into the larger 3-sphere because the nomally repelling forces are reduced. These condensates of spacetimes become as black hole seeds, and growth beyond a critical size or "critical mass", can indeed bring a chain reaction which consumes all matter within range of the condensates event horizon. We are positing here that that a black hole geometry is analogous to a giant vibrating atom, where the spacetime inside of a black hole has 3 non-Euclidean spatial dimensions and 1 time looping or frequency temporal dimension made of CTCs.

How can just empty spacetime ever manifest the characteristic of a physical solid?

Newton's laws of motion stipulate that anything moving will continue to move until acted upon by an external force. This is the law that has governed motion and it is permanently etched in our minds. So much so, that it is difficult to comprehend that nothing, **empty space, empty nothing** spacetime itself, can ever repel or deflect an object, and especially at a very discrete boundary. But if we have a **curved** empty nothing, then we have a gravity force through curved spacetime. GR proved that curvature in spacetime itself is enough to explain the force of gravity. It is the time dimension progression at rate C in spacetime that starts an object accelerating in a gravitational field, warped space. Moreover, it is the time dimension in elliptical spacetime when an object penetrates an atom boundary the differences in time components will produce and internal repulsive gravity force, ie a perceived collision and hence the very definition of a solid. Since time is frequency in an atom elliptic spacetime zone, anything in that spatial region will be undergoing a constant eigenstate expansion-contraction from one modeshape to the next. Hence anything entering that region of spacetime will eventually experience an expanding and enveloping atom spatial boundary with a different time component, the change in time effects are real physical acceleration experienced by the encroaching object, which will give the perception that the impinging object was repelled as the time component results in "spinning" space coordinates, giving the encroached object a deflected trajectory. The addition of an expanding spacetime grid would give the appearance of more space between the elliptical region and the impinging object. Since more space indicates a larger separation, we perceive that as a repelling from a "solid" atom cross section.

Like a spherical time vortex, anything physically contacting its edge or boundary undergoes a rotation or acceleration to an alternate trajectory because of the difference in the time variable. But different from an fluidic vortex, a time vortex operates strictly on the time component and space component, rotating the encroaching entity to a new time and space, which manifests as a physical acceleration onto a new trajectory or worldline.

There is also a spacetime effect promoting the physical "solidness" of matter perception, at an elliptical-geometry boundary defined region. Since time is circulating inside the atom volume of spacetime, the space continuum must compensate to remain uni-directional upon release or transition from inside the atom EGSFT into hyperbolic geometry space region outside the atom boundary. A "colliding" object meets no resistance to a physical boundary made of nothing more than something analogous to a physical shock wall or jump with different states of the same matter on opposite sides. An outside the atom object encroachment into the atomic boundary spacetime adds the new spacetime coordinates to it's local position, a considerable change to it's previous spacetime position. In addition the rotation or atom boundary spin of space due to frequency-time changes the physical coordinates to the incoming objects motion as well. Thus any object encroaching on a spinning spacetime spherical vortex will experience a change in velocity vector as its time component much change, "causing" an apparent physical repulsion or deflection, depending on the time-space of the entry or "collision", from the changing time values and space rotation transitioning from HGSMT to EGSFT atom space.

    Atom boundary spacetime sheets or membranes generally have effective stress-density and thickness for vibration in the spatial dimensions and temporal dimensions. Just as gravity is nothing more than curvature in space and spacetime subject to hyperbolic geometry outside the atom, so are the other "forces" such as electromagnetism, strong force and weak force, on progressively smaller dimensional scales and in non Euclidean geometries. We posit that all four known forces and the yet postulated fields such as Higgs, are nothing more than curvature of space-space and spacetime in non-

Euclidean geometry frequency-time spacetime and transitions of these as they map from the rules of one geometry, and time, to another. Stable time-space boundary is a separate topic (not treated in depth here). The boundary thickness is the spacetime vibration component, analogous to the viscous boundary layer between a fluid and solid. The spacetime "viscosity" is the dynamic attraction that occurs between the forever changing modeshape moving the atom boundaries between and around interacting atoms, molecules, and subatomics, almost just a mathematic concept infinitely thin, but having the affect of slowing their motion we perceive as inertia.

    Moreover, the boundary as geometric spacetime curvature outside the atom causes attractive forces outside in continuum with surrounding objects. With space or time or both, we will treat them seriatim, space-space, time-space, time-time.

    How the spacetime transformation occurs from one spacetime to another spacetime informs our research and application of this technology and indeed put us on to the road to reality from the abstract alone.

What's the Big Deal with Disparate Spacetimes?

    So what is the big deal with 4D submanifold spacetimes embedded in other 4D manifold spacetimes and the transitioning between geometries from Euclidean to Hyperbolic or Elliptic and back in the vicinity of atoms and molecules? The short answer is disparate spacetimes have disparate curvatures and these all are measured currently as mostly flat Euclidian except when dealing with large mass bodies. So, all our computer models and calculations for atomic and molecular structures are incorrect or flawed in fundamental ways because embedded spacetimes warp the embedding spacetime for very tiny almost massless entities. The big deal is that motions of atoms and molecules, vibration-rotation-translations, interact with each other over essentially straight lines, and straight lines are typically longer in Hyperbolic geometries and are curved, spacetime warpage. Just like the starlight beam measured going near the Sun during an eclipse showed the star's position was not where we thought because the beam was bent in spacetime around the Sun. The vibration – rotation – translation motions are how and where the atoms temperature or "kinetic energy" is accounted for. Since the actual vibration – rotation – translation motions comprise the kinetic energy for atoms and molecules, it is associated with and as a consequence of the loss of potential energy. Potential energy is one of position, the distance between the moving atoms or molecules must be known along straight lines. And straight-line distances must then be calculated to a certainty, where the distance depends on the geometry of the spacetime of occurrence. The current assumption that the Euclidean metric is the one to use in calculations is a basic error that can render calculations and atomic/molecular models worthless or at least provide erroneous measurements. The "measurements" then for the basis of the theories propounded to us are flawed.

Certain metamaterials designed with precise shape, geometry, size, orientation, and arrangement give them their smart properties capable of manipulating EM waves: by blocking, absorbing, enhancing, or bending waves, to achieve benefits that go beyond what is possible with conventional materials. These materials, negative-index metamaterials, are given a negative index of refraction which bends spacetime like a prism bends EM but through multiple tiny lenses and prism orientations which bending the geometry of spacetime using composite material orientation repetitive but not uniform fashion.

Where non-Euclidean geometry of reality becomes vital, Solid State Physics, Particle Physics, Chemistry, Material Science, and Metamaterials will be the primary areas that can quickly benefit from Chronature models of disparate spacetimes and interactions between them.

Paraphrasing Einstein on GR, "matter tells spacetime how to curve and spacetime tells matter how to move". Chronature model reverse escalates this concept to the molecular level where the atoms and molecules tell spacetime how to curve straight lines and spacetime tells atoms and molecules how to vibrate-rotate-translate. and where Maxwell's EM model is simply the transition from one spacetime to another.

That atoms and molecules act to attract each other is observed and known. As magnets, gravity, or electrical charge, the spacetime curvature inside and outside of an atom creates attractive and repulsive forces. The forces are arising from the same effect, different spacetime configurations with curvature manifesting the forces that we perceive and measure. The question has always been why and how does this happen physically, or as we shall see geometrically? How and why interatomic and molecular forces do this is currently speculation that has lead to 50+ theories of bonding, based mostly on empirical evidence for very small windows of state and phase. The differences in the strength of attraction are how the forces of attraction are traditionally classified and named. Observations show proximate neighbor atoms have lower "potential energy" at certain close distances from each other but if they get *too* close, they start repelling, also currently unknown, or unknowable from an electron orbital model.

    Where there is but a single atom in Minkowski spacetime, a background surrounding field is produced as shown about, a field that distorts the surrounding space. A test particle is one such entity whose own field is negligible and does not affect an atom's surface acceleration field, so it does not change the atom's surface acceleration field. But an atom in another atom's field, molecules and matter, change each other's fields, and this is currently studied as "bonds". There seem to be as many "bonds" as elements and compounds are comprising these.

    We now enter into the realm of Geodesic Deviation. Since at the matter molecular level, we have a more general space, the "straightest possible lines" are geodesics. So in the acceleration fields created by the atomic boundary curving space, atoms will exhibit shared accelerations between them which cause Geodesic Deviation in the field between them.

As atoms away from each other may start in flat space, the geometry quickly complicates with connecting vectors from one atom boundary geodesic to another atom boundary geodesic at the surfaces. These are currently handled by the General Relativity tensor calculus, covariant derivatives, partial derivatives of the connecting space coordinates, and Christoffel functions to handle the deviations. This is unexplored territory and a result of the current Chronature model. Suffice it to say that currently these atom and molecule dynamics have rudimentary models of particles undergoing rotation, translation, and vibration and these models constrain us to the use of statistical thermodynamic models of large populations for engineering empirical material properties.

Atoms having low "kinetic energy", below certain threshold translation – vibration – rotation motions, do not have sufficient motion to "break away" from the space warping attraction called a bond, continue to go back and forth in the immediate local vicinity relative to each other. The geometric distance among neighbors and the degree of freedom from attractedness dictates the state of matter that we see: solid, liquid, gas, plasma, or condensed.

The reason why state or phase transitions are discrete is atoms/molecules act to balance local attractive and repulsive effects of spacetime curvature dynamics. To state it simply is known, the reason is that the molecules in the phases behave differently, specifically in how they move and attract and repel neighbors in the spacetime geometry that they create about them and interact with other atomic surface spacetime accelerations. Atoms in a solid can vibrate back and forth and rotate through the hyperbolic geodesic or "straight line" connecting atoms in a lattice structure, but the molecules in a solid cannot bend and flex to the extent that they can in a liquid configuration, generally, because the individual molecule kinetic fluidic flexibility from interacting local spacetime acceleration with and generally non-Euclidian geodesic length distances between sites. In the liquid and gas phases, the molecules can have translational kinetic energy, that is, moving in a hyperbolic geometry "straight line", which those in a solid do not - as they only move back-and-forth, along a hyperbolic straight line between atoms. Molecules in the gas phase do not have intermolecular bonds that can vibrate and therefore cannot store kinetic energy in vibration because neighboring position sites, back and forth along a geodesic, so their kinetic energy consists of translation and rotation. The figures below show the "straight" lines between atoms in our current models and the actual "straight" lines between atoms due to the hyperbolic geometry of spacetime in the vicinity of the atoms.

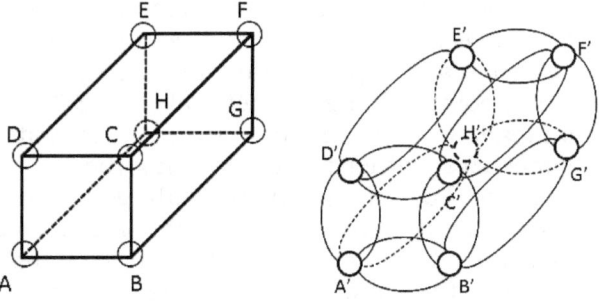

Please imagine all of this happening with spacetime "bubbles". The curvature of the spacetime bubble varies with time, that is the time on the inside of the bubble changes the bubble into its fundamental modeshapes but so quickly that on average it's in the general shape of a sphere. But in the proximity of another spacetime bubble that has similar modeshapes within its spacetime borders, both will "lock on" to each other's dominant modeshapes in resonance, forming a "covalent" bond.

Now picture just two-atom "bubbles" attached by a warped space, this would be a warped line revolved around a straight line center axis connecting two spheres shown below.

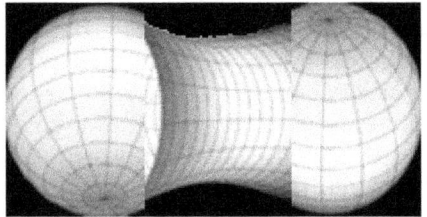

These differences in how the "energy" or motion between atoms is distributed are ultimately what cause discrete phase transitions. At a phase transition point, adding geometrical "energy" or motion leads to those motions being redistributed or re-geometrized, rather than increasing the temperature. For example upon heating ice will increase its temperature, atom vibration and rotation increases, until it reaches 0 C, then state change - melt without the temperature increasing. This means that the kinetic energy goes to translation and rotation - a vibration of the water molecule freeing up the solid ice until all is converted to liquid. Once the ice is melted increasing the translational energy of the liquid molecules will then increase its temperature, because the molecular vibration and rotation will increase until the liquid water reaches 100°C for the remaining ice molecules to completely unbind from the molecular attraction spacetime curvature between molecules. Continuing a temperature increase will find subsequently "temperature" stays steady at 100°C until all of the liquid turns to vapor, as temperature measurement only registers collisions of translation density. Once the kinetic energy of the molecule translation to all the liquid water molecules turns to the gas collisions, we find increasing gas translations register an increasing temperature again, turning to superheated steam.

   Phase transitions are unique to the atoms and molecule formation because the spacetime between them is unique, depending on the curvature of each entity and density. For example physically compressing steam, pushing the steam into a smaller volume to increase the pressure sufficiently, will push the gas molecules together into a liquid, which changes spacetime curvature into vibration to rotation, and different modeshapes harmonic resonance at alternate steady state band angles, then "stores" the kinetic energy in the proximity of the molecules, water bond. As atoms and molecules get "pushed" together, lines of spacetime must follow the atomic boundaries preserving the atom spacetimes, and the geometry immediately outside the atoms becomes more eccentric, i.e. straight lines have larger arcs and actual distances between atoms are increased, increasing potential energy. But if the temperature is increased so that all the molecules have enough "energy" that they'd quickly gain the kinetic motion of translation, then a solid can transition to gas directly instead of going to the liquid phase first because of the "pent-up" potential energy in spacetime geometry.

At the critical point, below which there ceases to be another phase transition, the matter state becomes continuous. In that region all the kinetic atom and molecular motions exist, displaying neither a liquid nor a gas, but a dense liquid having a very low viscosity like a gas, viscosity being the property of attraction at the molecular level. This is a supercritical liquid or gas. Where the transition of the attractive force to lower more fundamental modeshape resonances and proximities between atom sites. Whether from the temperature or pressure, atomic and molecular kinetic energy, a supercritical fluid may transition to liquid or gas without a discrete phase transition. Thus state transitions from liquid to gas without any discrete transition happen via a supercritical fluid.

In a solid-state, an element or compound can be in a crystal or semi-crystal structure. This means that in non-Euclidian space in the local vicinity of atoms, the atoms have some rotational and vibrational but limited translational freedom. Moreover, not only the modeshape or space curvature but the individual atom harmonic frequencies can add unexpected properties. For example at super-low temperatures, a crystal called samarium hexaboride behaves in an unexplained current model-defying way. This much-studied compound, samarium hexaboride or SmB_6, is an insulator at very low temperatures, but researchers in Cambridge observed electrons traversing orbits millions of atoms in diameter inside the crystal in response to a magnetic field; mobility that is only expected in materials that conduct electricity. An experimental condensed matter physicist at the University of Cambridge said the discoveries she and her colleagues have made "mean that something needs to be rewritten completely."

Moving to a liquid state, a crystal structure is disrupted and the individual parts of the bulk have an unfettered motion to rotation and vibration but limited translation because the translation motion causes attraction from more than just primary immediate neighbors. These are currently articulated and characterize as interactions between the materials from either charge in the case of salt, or dipole-dipole interactions in the case of polar compounds, or induced dipole interactions in the case of non-polar compounds. With current models, even this is almost impossible to numerically predict as the models are incapable.

In a gas form, molecular translation is unconstrained, molecules separate from the bulk population spacetime with bond freeing kinetic trajectories which randomly lead to dispersion. Each molecule or atom is free to effuse into free a flatter space or diffuse into matter which provides attractive bond curvature. Diffusion into matter is more random but rates depend on types of attraction from the geometry of the matter spacetime boundary and the geometry of the spacetime in between. But atomic or molecular rotation- vibration- translation is difficult to model unless one can predict the movement path, and movements in hyperbolic space curved spacetime occur along straight lines that are "curved" as shown distinguished in the ideal perovskite structures show below in Euclidean and non-Euclidean Hyperbolic geometry space.

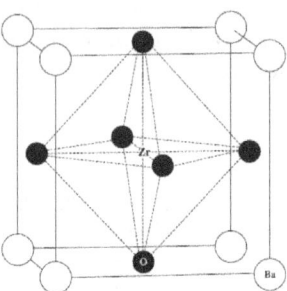
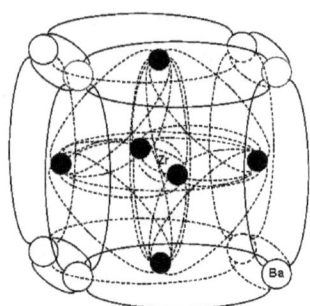

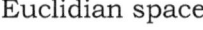

Euclidian space　　　Hyperbolic space

The lattice in the Euclidean space is shown on the left as the is the classical representation of the solid structure. However, a Chronature representation, because the proximity of atoms demands a hyperbolic spacetime for straight lines each atomic site in a cubic structure with vertexes in hyperbolic space showing a representative two lines connecting it with each neighbor. These are better represented above by a tunnel surface of revolution between the atom spacetimes. The two lines represent an infinite number of straight lines between the neighbors as in Hyperbolic Geometry at points in curved spacetime which can have an infinite number of parallel line connectin atom sites, and not just one straight line as would be the case in Euclidean geometry space. The 90° angles between neighbors can be closer to 180° angles in non-Euclidean space. Moreover, a translation between more than immediate neighbors can be a straight line because orthogonally has a different physical appearance and meaning, a geodesic, in curved space. Therefore determining the "energy" of rotation, vibration, or translation in Hyperbolic spacetime models will be vastly different in each geometry, and using the incorrect geometry would yield erroneous results. In addition, the actual geometry curvature would change as the atoms and molecules changed position. So it's the actual dynamic geometry and the geometric configuration of the atoms/molecules that determine the phase state. So spacetime geometry tells the atoms where to locate, and atom locations tell geometry how to configure the straight lines between them.

We bring up Perosvskovite here for a reason because although we have had Perovskite since the 1800s, Perovskite is only recently found useful in a type of solar cell material, for the light-harvesting active layer. Perovskite materials, such as methylammonium lead halides and all-inorganic cesium lead halide, are cheap to produce and simple to manufacture.

Solar cell efficiencies of devices using these materials have yielded as high as 28.0%, exceeding the maximum efficiency achieved in single-junction silicon solar cells. Perovskite solar cells are therefore the fastest-advancing solar technology to date, with the potential of achieving even higher efficiencies and very low production costs. However, uses are all being discovered by accident or trial and error because although there are mathematical models of the crystal structure based on what is currently known about atoms and lattices, measurements of "bandgap" and other useful known properties, no solid-state models of matter exist that can predict these kinds of materials, or how or why they work. Perovskite is a calcium titanium oxide mineral composed of calcium titanate ($CaTiO3$). The shell structure for Perovskite as per the periodic table

Ca   [Ar] $4S^2$
Ti   [Ar] $4S^{3+} 3D^2$
O   [Ar] $1S^2 + 2S^2 + 1P^4$

The atom modeshapes of these elements having these harmonics will define the geometry of the spacetime surface curvature of the atomic surface configurations for the lattice structure.

Are there other geometries/ topologies such as discrete geometries that can be governing in the solid phase lattice structures, where the movement has limitations as to among neighbor atoms/molecule sites?

Can't we just start with some discrete space like a lattice structure, provide a metric and do all the modeling mathematical work for it in the discrete space? Can't the discrete lattice structure then "at large length scales" have this underlying lattice reduce to a Euclidean geometry? How does the lattice add to distances as between sites, and can atoms/molecules of two sites swap sites without a third site to act as the temporary intermediary location, ie can they "commutative property" in real physical space?

Without getting deep into the holographic principle, amplituhedron, uncertainty, or positivity, we believe that motion between sites in a lattice structure does not require discreteness but must be addressed in the computational models for these aspects. At first blush, one might think a simple straight forward approach would be to implement a 3D rectilinear Euclidean distance metric, similar to the 2D taxicab metric modeling positions separated by city blocks, the grid layout of most city streets, and the shortest path a car could navigate between two intersections. This would amount to the shortest path length sum of the actual lattice site legs from start to finish, as shown in the diagram below- showing Taxicab[26] geometry versus Euclidean distance.

---

[26] *http://en.wikipedia.org/wiki/Manhattan_distance*

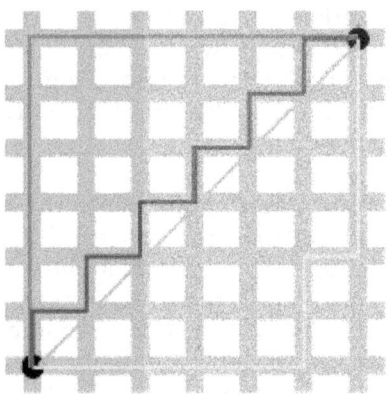

In taxicab geometry, all three shortest path lines shown have the same length (12) for the same start-finish intersections. In Euclidean geometry, the shortest path distance has a length approximately equal to 8.5 and is the unique shortest path. Imagine this in the Hyperbolic Geometry lattice shown above with geodesic or "curved" straight lines between sites would give yet another metric. Using a kind of compound model in the non-trivial modeling implementation of the solid phase lattice structure of solids integrating the geodesic distances would give a vastly different result and understanding.

In current GR, the Minkowski space is flat with only a one-time dimension typically plotted against multiple spatial dimensions. Because each atom is its spacetime, a Lorentzian manifold may be better used for the metric tensor. The Lorentzian would model space for a pseudo-Riemannian manifold of signature (p,q) is $R^{p,q}$ where the p and q are for multiple space and multiple time dimensions, as each atom would contain 3 space and 1-time dimensions.

Where does the Plasma state fall in all this?

The plasma is the 4th state of matter and demonstrates again that each state of matter has a unique set of fundamental freedoms of motion in spacetime or restrictions that other states do not and which we perceive as characteristically and physically different as a group. In traditional states, ions, atoms, or molecules with geometric shapes are attractive to other atomic entities. These are bound by other attractors, electrostatic interaction, to each other. In plasma state, sufficient kinetic motion is required to break these "ionic" attractions as governed by spacetime, dynamic spacetime curvature interactions. Naked ions, atoms with open harmonic configurations are traveling in the gas phase, unbound as a counter ion, attracting opposite spacetime curvature atomic entities.

In between the liquid to gas states of matter, there is a spectrum known as supercritical fluids. Supercritical fluids are formed where a liquid atom/molecule is imparted kinetic energy but under a volumetric constraint, under pressure. Basically, the supercritical fluid translation motions increase to fill its volume constraints, but the intermolecular attractions are still significant, and so it has characteristics of both liquid and gas states. The big benefit in studying this near-zero matter is that the spacetime curvatures are less dynamic, and an absolute zero is much easier on calculations of spacetime curvatures with inter-atom spacetime interactions offering significant differences in properties of interaction results.

How does this affect atom "bonding"?

Bonding effects can be modeled from a first simple linear very rough approximation approach accounting for the geometric effects as:

Bond Strength =

(1/ (distance) ^{2}) * (SpaceCurvature at boundary) * (ShapeFactor)

Through the judicial use of scalar spacetime curvature, gravitationally affecting curvature areas, and superposition of shape factors, we should be able to create reasonable bonding models.

Of course, it will depend on what kind of a "bond" it is. For example what if the bond is an entanglement of atomic spacetimes through their temporal umbilicals as depicted below?

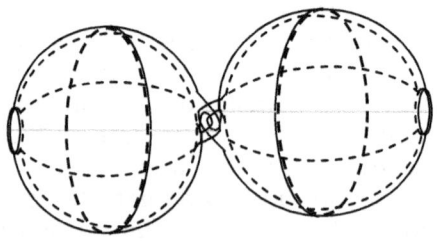

The above configuration of two atoms may manifest physically as some kind of magnetic alignment or even a magnetic property of the matter, the smallest granularity of a "domain". On another and opposite size scale, the black hole Quantum Entanglement is held in QMT theory is to be the building blocks of worms holes, where great distances in space can be connected through an Einstein-Rosen bridge. This intergalactic region of spacetime is not the target of Chronature. However, a black hole solution to the EFE may have atomic spacetime on that very grand scale having frequency-time and 3 non-Euclidean Elliptical space dimensions.

How do Chronature models enable us to calculate inter atom and intermolecular forces and reactions where QM fails us?

Because the state of the art in computational chemistry uses primitive Bohr-QM in Euclidean space models, it's very difficult to impossible to make calculations reconcile with known experimental data as the Quantum state space becomes enormous and cannot be computationally modeled, even with supercomputers. Therefore predicting the result of introducing given reactants for interaction, is usually done via by-guess-and-by-gosh. Hence an experiment is designed and results measured in a simpler easier way. Then a theory can be developed to explain the results.

Although the building of new compounds in computational models is done with some regularity, the results are not determinative or definitive. Here, time becomes a synthetic chemist's most valuable fudge tool. Without a timely answer, they are free to speculate. With the current models and tools, if it can at least show with calculations that a given molecule with the desired features isn't impossible, then it's probably worth trying to make the synthetic. Computational chemists inform us that some of the most important and interesting features of molecules are the hardest to model with computer calculations. The extremes of bonding and reactivity, very weak interactions or bonds on their way towards moving on, are exceedingly difficult to model correctly, and predictions that invoke these states are often inaccurate. This is despite the computational chemists' perceptions that we can correctly model nearly any molecule that we can imagine.

Chemists claim that the physics governing chemistry has been known for nearly a hundred years and it is the only implementation that is difficult or impracticable. Moreover, some chemists contend that there are computational methods that allow them to access a high level of accuracy called full-CI, the computational equivalent of a full-Monte, but the computing time increases hopelessly with the factorial of the number of electrons in the atom. This precludes using current computational chemistry methods on a serious protein anytime in the foreseeable future.

At its computational modeling foundation, chemistry turns to QM. Initial calculations are from QM precepts, and even modeling the orbital of a single hydrogen atom, starting from the simplest but no simpler, one quickly discovers how precarious it is to properly and truly simulate chemistry numerically with computer models.

The QM calculations involved become horrendously complex. Computer chemists must then as a last resort, resort to modeling shortcuts and approximations. And to nobody's surprise, these tend to mess up the results.

Attempting to approximate even a two-electron orbital atom can be futile. Hydrogen reactions are difficult but they can be done analytically. "Electron-electron", fundamental harmonic modeshape interactions worked out on papyrus and charcoal straight up becomes impracticable. And that's still only a single orbital. Don't attempt to resolve the d or f orbital without a ton of computational horsepower.

It's a bit easier to use a real atom and call it a perfect analog simulation of an atom. But upon determining what state the atom is in, one must change the simulation. Strange as it sounds, with current computational models the best logic we can apply is when modeling a quantum mechanical system, use a "quantum mechanical system." This forces the most computationally difficult tasks to just take care of themselves because the analog simulation is much faster than digital.

The Chronature models outside of the atom are only a matter of Hyperbolic non-Euclidean geometry turning to Euclidean Geometry outside the $1/R^2$ distance of acceptable approximation. Inside of the atom we have Elliptical non-Euclidean Geometry with subatomic nucleon particles bouncing against the "strong force" emanating from positive curvature inside spacetime repelling towards the atom center. Concurrently the nucleon mass exerts an attractive gravity on the atom boundary wall. A point mass model of the nucleus is a very accurate starting place for these internal force calculations as well.

How does Gödel's closed timelike curve (CTS) fit into frequency or circular time Chronature?

It's difficult enough to wrap one's head about moving through time as we do along a world line. But rotating in time is probably a much larger mind-bender. But in short, Gödel's CTCs and "swirling dust" or "rotating universe" metric form the mathematical physics basis and foundation for Chronature frequency or circular time or the dimension of time in atomic spacetime. Time loops, or circular time, have been mathematical speculation by **Gödel and others**. A CTC or time loop is a world line in a Lorentzian manifold, of a material particle in spacetime, that is "closed", that is a worldline returning to its starting point. This coincides with the Chronature model of an atom vibrating sphere spacetime. Solutions to the equations of GR allowing CTCs, known as the **Gödel** metric and other GR solutions to Einstein's Field Equation containing CTCs were found by **Gödel** but were never intended to model the atomic spacetime. So we are free to speculate that CTCs exist in actual physical form but inside of atoms and certain subatomic particles as well, supporting the Chronature model of physical atoms each with their spacetime, where time is a CTC or frequency-time for their spacetime time dimension. CTCs opens a wide area of speculation even further, and for our purposes, we will stay close to the simple atomic model vibrating sphere with 3 space and 1 frequency-time spacetime. Unlike what was envisioned by **Gödel for a local universe that was time rotating,** in Chronature the **Gödel metric tensor** becomes an embedded submanifold in macro spacetime where -when we live. Moreover, this also provides for more than the one-time dimension and different time dimensions, something which is currently constraining the String Theories metric tensors to one time and multiple spatial dimensions.

3 – Atomic Structure

How is it that we know the radius of a given atom or element when their "electrons" are busy orbiting the nucleus? The answer is we don't really; the atomic radius is an average distance from the atomic undulating boundary surrounding a nucleus. Given the fact that only at absolute zero is there no atomic vibration-rotation-translation to interfere with the measurement, the measured atomic radius must be only a kind of average or mean distance from side to side, outer "shell electron" of the surrounding "cloud of electrons." Because the physical model of the atom is confused, ie QM explains that the electrons do not have orbits or sharply defined ranges, their positions must be described as probability distributions that taper off gradually from their actual distance from the nucleus. What the "ground state electron orbit" corresponds to physically has always been in question since QM is not a real physical model; many physicists have given up using physical models and only probabilities of states of possibilities are required for QM's non-physical models. It's anybody's guess what the current atomic radius measurements truly are off, but we can speculate as well that they are averages of some sort.

Bohr's model, only good for Hydrogen, suggests that 1) the orbit nearest to the nucleus is lowest in "energy," 2) when an electron "drops" to a lower orbit it emits a photon, and 3) when an electron absorbs energy, it "jumps" to a higher orbit.

These are all a bit like Newtonian mechanics for celestial bodies' pre-GR. However, Hydrogen is the limit for the orbital mechanic model since it has only one orbiting satellite and another would mean we would need the solution for a three-body problem.

There is no general analytical solution to the three-body problem given by simple algebraic expressions and integrals. Moreover, the motion of three bodies is generally non-repeating, except in special cases too simple to model reality for most cases. Using a computer, the problem may be solved to arbitrarily high precision applying numerical integration although high precision and n bodies require enormous amounts of CPU time. Hence even in QM, the atomic hydrogen orbital model was quickly discarded as not usable for elements having more than 1 orbiting electron, Hydrogen.

Under the present best measurements the radii of isolated neutral atoms range between 0.5 and 3 angstroms. Therefore, the radius of an atom is more than 10,000 times the radius of its nucleus, by Euclidean metric, and less than 1/1000 of the wavelength of visible light (4000–7000 Angstroms). But since these measurements were made in the three macro spatial dimensions Euclidean coordinate system, not inside the atom, they are suspect for accuracy inside the atomic sphere boundary, with a completely different and non-Euclidean coordinate system. Under non-Euclidean elliptical geometry, Riemannian manifold space, this would measure differently as the spatial dimensions inside the atom is not uniform or "flat", Euclidian, as they are currently measured outside of the atom, see figure below for an example of uniformly increasing dimension bands inside a Euclidean vs non-Euclidean space. All bands are of equal width; to reveal the nonuniformity of geometries.

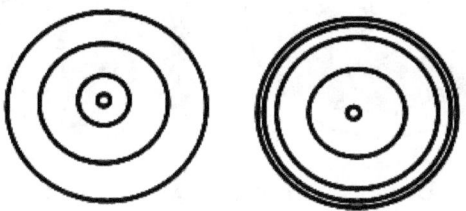

That is to say, where units of a dimension metric remain uniform or "shrink" as they approach larger geometrical values, changes the nature of the objects within that spacetime. Most likely a dimensional mapping change will help us to understand what cyclical time and elliptical space look like and behave. This will give us a better understanding of the actual structure of matter.

Apart from that, Escher does a pretty good job of illustrating the elliptical geometry below; the left shows a Euclidian space while the figure on the right is drawn in non-Euclidian elliptical space. As the birds and fish in the Euclidean space are all the same size end-to-end, the angels and demons in the figure on the right are not so clearly the same size, because the geometry gives them a shrinking appearance toward the outer edge boundary, since the lines of space get shorter as we go outward radially: or conversely, long as we approach the center. See the distinction with the below Eschers.

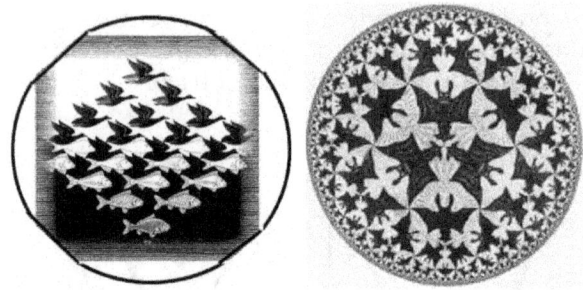

Under current element classification many characteristics of the elements are associated and explained by the position, row-column, that they hold in the Periodic Table in accordance to their atomic number and electron shell structure under as predicted under Schrodinger derived, electron shell, model, purportedly giving rise to their chemical, electrical properties, and all other attributes.

Early on it was thought that all atoms of the same element have the same radii. However, there are many ways to measure atomic radii, and they have different names for different measurements, including Van Der Walls radius, ionic radius, covalent radius, and atomic radius. Even so, it was determined that an atom does not have just one size, because the bond determinative modeshapes will dictate atomic radius as measured. And lest we proffer that it is an insignificant assumption, such imprecision makes synthesis and construction of innovation based on these inaccuracies even more difficult to manage, much less understand or predict the workings or results. The atomic radius is vital to understanding where the discrete changes in curvature occur, a parameter that is vital in predicting precise interactions between atoms. That said we can go a long way to understand curvature by using the inverse of the average radius squared for some purposes.

With more information and measurements came atomic and chemical properties and consequently, more theories were needed to explain them. Hence we have "shielding" defined as, "attractive force acting on electrons by protons in a nucleus, a repulsive force acting on outermost shell electrons by inner electrons, Lanthanide contraction, D-block contraction" and so forth and so on and on.

The Chronature physical model of the atom is not composed of electrons in probability clouds, but a 3-sphere comprised of spacetime curvature boundary, the physical boundary between at least two spacetimes. This 3-sphere atom model has a one surface boundary that interacts with the spacetime outside of it, and is represented as a spherical wave with the classical spherical wave equation for radial dimension in time as:

$$\Rightarrow \nabla^2 p - \frac{1}{c^2}\frac{\partial^2(p)}{\partial t^2} = 0 \Rightarrow \frac{1}{r^2}\frac{\partial}{\partial r}\left(r^2 \frac{\partial p}{\partial r}\right) - \frac{1}{c^2}\frac{\partial^2(p)}{\partial t^2} = 0$$

Thus the boundary of the atom shown in the checkered modeshape solutions below subtends another kind of spacetime, and its curvature magnitude in space and time modeshapes provide the "chemical" and "electrical" properties that we see and measure. Because they are spacetime curvature at specific times, the spherical harmonics determine the atomic boundary curvature, which in turn, determines the atom's "chemical" and "electrical" properties, at specific times.

The spacetime outside the vibrating atomic 3-sphere boundary, negative curvature near the boundary, and the spacetime inside the vibrating atomic 3-sphere boundary, with positive curvature on the inside. Since the degree of curvature determines the force as we have seen from GR, the outside curvature of this boundary determines the atom's chemical and electrical properties, as these are the forces from warpage in the vicinity of spacetime. Furthermore, curvature depends on the atom's harmonics and modeshapes to give it the curvature of attraction or repulsion that produces the properties of each element.

What makes the Chronature atom model 3-sphere vibrate at the electron structure harmonics whose modesphapes define the spacetime curvature?

The short answer is that some particles inside of the atom sphere are causing the vibration at the natural frequency of spherical harmonics. Why is that, certainly the fact that atoms themselves are banging away at each other, vibrating, rotating, and translating in space, should be enough to cause sphere vibration? But these are random impacts from attractive atom boundary curvature caused fields that promote the overall motion of a population of atoms producing alternate states of matter more or less proportional to the overall population's activity.

Hence, the answer lies within the atom. As we will proffer below, a proton, it's spacetime own spacetime composition to be shown later, is a charged "particle" that lives within an atom's spacetime. The atom wall has a mathematical closed-form solution spacetime curvature and that spacetime curvature creates a repulsive field inside, as opposed to an attractive field outside For the moment, let's not worry about what kind of field, except that the field produced by the curvature inside is called the Strong Force and is repulsive, not unlike one of a magnet's poles. The Strong Force (SF) repulsive field inside an atom pushes any particle towards the atom center. This is where it gets interesting because as Wheeler observed[27], "Fields tell charges how to move, charges tell fields how to vary." And so the charged proton tells the atom wall to vary. And it does so according to Hoyle, by the math cards dealt in sphere vibration. Never mind what charge is or represents for the moment. So, for every proton, a spherical harmonic is added to the atom wall. We can analogize this by adding a mass into a model for each proton, attached to springs, representing the actual physical atom effective wall flexible properties, attached forcing functions from the inside of a sphere wall. The masses m_1, m_2, ...m_n, for each proton, will impart a system natural frequency of f_1, f_2, ...f_n respectively via the springs k_1, k_2, ...k_n. The atom boundary has a physical thickness and density which we can see through our AFT microscopes. This atom boundary has flexibility and is not unlike an object itself, having at least two

---

[27] Leonard Susskind, "Special Relativity and Classical Field Theory, The Theoretical Minimum", pg 155

fields pushing and pulling on the boundary. Visualize a balloon with marbles bouncing about inside causing balloon surface vibrations. Consult the figure below for a mechanical model analogy.

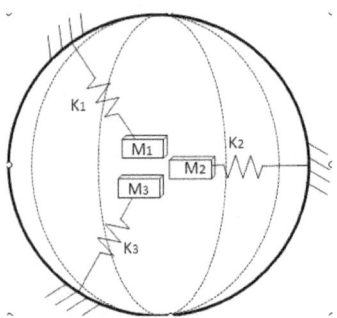

On the topic of how fields, spacetime curvature of the atom boundary, affect particles, nucleons, and particles affect fields, consult Susskind on a great explanation beginning with the fundaments using the Euler-Lagrange equation. Susskind, probably the closest teacher of physics we have to a Feynman today, begins with the Lagrangian action principle using geometry alone, for an elegant and easy to understand approach with the application of various calculus principles, where the sum of the potential and kinetic components of a particle in a field have action and reaction together in the same event which governs the motion of particles immersed or embedded within fields. Regarding fields and particles in electrodynamics, Susskind explains it this way, " The fact that the field affects the particle tells us that the particle affects the field."

This may feel unsatisfying to some because we are talking about the interaction of curved spacetime, forming two fields with an alternate curved spacetime boundary, the atom wall, which by virtue of its spacetime geometry of curvature projects one field inside the atom pushing particles towards the atom center and another field attractive to outside atom boundary particles. Said another way, curved nothing interacting with curved nothing of a slightly different spacetime curvature geometry produces a separation in the form of a boundary that causes a repulsive field toward the curved nothing's center. While this is pure geometry of spacetime it nevertheless, through the mere passage of time, causes the physical phenomena that we experience as matter having its properties. It may be easier to understand through a fluid mechanics analogy, whereby a whirlpool spins in a body of water, and the whirlpool is spiraling fluid into its center. The water and the whirlpool are the same substance, water. At some distance from the whirl, the pool center defines a vortex, where the fluid outside the vortex is not engaged with the whirlpool mechanics. That outside will be the ambient river, lake, or other body of water. The whirlpool vortex of a cylindrical 3D volume but analogous to the 3D atomic spacetime is pushed any floating objects into the whirlpool spiral center. It's not more than moving and rotating substances are the same or like substance, yet they are separate and represented as different fields where the fields will affect "floating" objects differently in the different fields of a stream. This is a physical analogy of the real and physical atomic structure and interaction.

The following diagram shows any number of asymmetrical shaps that the proton can assume to provide a charge. The charge occurs when particular points on a particle surface, are of different curvature magnitudes, asymmetric, and highest magnitude curvature of particle surface is attractive to that particular point because the net attraction is positive or negative. Wherein the same curvature surface would be attracted by all points on the surface equally, providing no net attraction to another particle in a field.

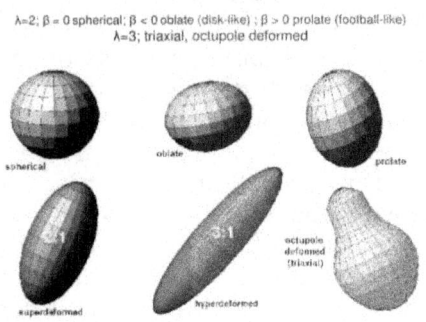

$R(\theta,\varphi) = R_0(1 + \beta Y_{\lambda\mu}(\theta,\varphi))$
λ=2; β = 0 spherical; β < 0 oblate (disk-like) ; β > 0 prolate (football-like)
λ=3; triaxial, octupole deformed

Enter, Spherical Harmonics, a geometrical property of vibrating spheres.

Regarding 2D spherical surfaces embedded in 3D, the Spherical Harmonics, $Y_{\ell,m}(\theta, \varphi)$ - solutions to the vibrating sphere equation, are functions defining a hollow sphere or atom boundary geometry. As was found for a Hydrogen atom, the advantage of spherical coordinates is that an orbital wave function is a product of three factors each dependent on a single coordinate: $\psi(r, \theta, \varphi) = R(r) \,\Theta(\theta)\, \Phi(\varphi)$. The angular factors of atomic orbital $\Theta(\theta)\, \Phi(\varphi)$ generate s, p, d, etc. functions as real combinations of spherical $Y_{\ell m}(\theta, \varphi)$ (where ℓ and m are called quantum numbers that are modeshape defining numbers to the vibrating sphere harmonic solutions.

These harmonics solutions to a spherical wave have certain constraints. One can imagine these as the surface undulations and oscillations on a soap bubble only much, much smaller and much faster. On constraints for example, if the volume of the bubble is constant, $Y_{0,0}$ is not used. If the center of mass of the bubble is constant, $Y_{1,m}$ is not used. The lowest frequency oscillations of a soap bubble are $\ell = 2$. The radius of the soap film is $r = 1 + \varepsilon\, Y_{2,m}(\theta, \varphi)$. The oscillations with different m all have the same frequency. The shape of the oscillations with m = 1 and m = 2 are the same up to a rotation, but the m = 0 oscillation is different. So we can see that the shape of the sphere changes by defining certain constraints as well.

Since the modeshapes of these harmonics are boundaries between spacetimes, they are spacetime curvature and define the curvature at any point on the atom's boundary at specific times or resonances. A few example modeshapes arising from solutions to the spherical wave equation are shown in the figure below. These do not show a degree of curvature, only that the spherical wave zero's and maximums as they exist on the sphere.

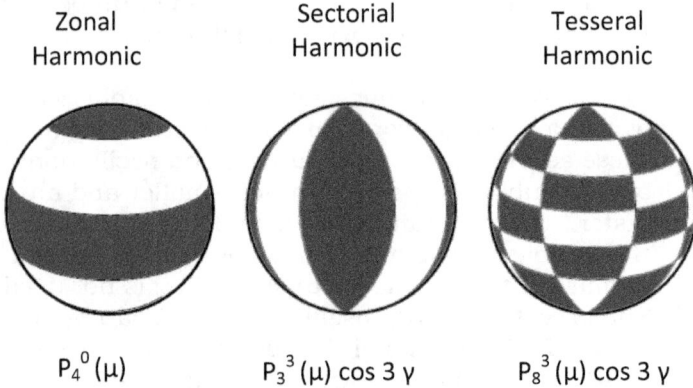

Zonal Harmonic	Sectorial Harmonic	Tesseral Harmonic
$P_4^0(\mu)$	$P_3^3(\mu)\cos 3\gamma$	$P_8^3(\mu)\cos 3\gamma$

When the spherical harmonic order m is zero, the spherical harmonic functions do not depend upon longitude and are referred to as zonal. When n = | m |, there are no zero crossings in latitude, and the functions are referred to as sectoral. For the other cases, the functions checker the sphere, and they are referred to as tesseral; Please note the natural symmetry from math to physics.

These purely mathematical solutions to a spherical vibrating shell provide us with calculable physical curvature. The actual atomic boundary curvature is determined by the perturbations, forcing functions, from nucleons inside the atom's boundary and collisions from outside the boundary also affecting radius. Inside "subatomic particles" moving about in the nucleus stimulate natural frequencies of the sphere to become excited with natural frequencies vibrating sphere per harmonic. These "subatomic particles" are currently measured as in Euclidean geometry spacetime as 10^{-5} smaller in radius. Euclidean measured outside the atomic spacetime give 3-spheres called protons, produce forcing functions from "mass", very high spacetime curvature entities, residing near the atom center, nucleus. These interact with the atom sphere boundary, causing vibrations. Protons are eccentric or asymmetric, not 3 dimensional symmetric 3-spheres are repelled from the atom inside boundary due to the atom inside the boundary is of positive curvature with gravity repelling forces. This constant bouncing back and forth, inside the atom, is what excites the spherical harmonics on the atom boundary creating surface harmonics, AKA "electrons". These give the atom sphere boundary its harmonics, its outside shape, and curvature which determines its chemical properties and molecular dynamics. The "neutrons" are also similar in radius to the protons but have complete 4D symmetry, thus not exciting any inside wall any more than another.

As stated above and quoted from reference, the QM model obtains the same electron shell solutions as the vibrating sphere for the harmonic modeshapes, shown here:

$\phi \rightarrow e^{im\phi}$

$\theta \rightarrow f_{lm}(\theta) \rightarrow$ Legendre Functions

$f_{lm}(\theta)g(\phi) \rightarrow Y_{lm}(\theta,\phi) \rightarrow$ Spherical Harmonics

$R_x(r) \rightarrow$ Laguerre polynomials

$\psi_{nlm}(r,\theta,\phi) = C_{nlm} R_{nl}(r) f_{lm}(\theta) g_m(\phi)$

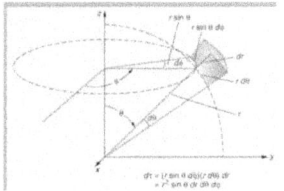

Because these figures are derived from the QMT model, which takes out the time for probability, we don't have a time dimension here. However, we can obtain the curvature of the space-space portion of the spacetime curvature, using the atom radius of

curvature for the different harmonic modeshapes, applying them in time dimension as averages or as individual surfaces where there is resonance. Since the time-space component is a small contributor to the total spacetime curvature, we can work independently of time at the atomic level with a good approximation.

As an example, the spherical harmonics in 3D are shown for the $P_n^m(\cos\theta)\exp(im\phi)$ harmonic of degree n=5, and orders m=5,4,3,2,1,0,-4,-5 below. More on the relationship between the **m,n variables, and s,p,d,f, etc "shells",** as we get into the vibrating sphere harmonics or as they are called quantum numbers. Naturally the tesser checkered spherical modeshape, with higher atomic number shells specifically d and f arise from higher spherical harmonics, higher element number atoms like the Y(8,4) shown below.

Although the shape of the atom is largely spherical, on the average, 3-sphere, it is only an average that our vibrating spacetime sphere atom makes over the timespace, where electrons are not orbiting particles but 3D spherical harmonics of spacetime curvature bounding the inside spacetime from outside spacetime of an atom. This is analogous to the Fourier harmonics which sum all of the harmonics for any given time to provide the curve position at that particular point in time. From these, computer simulations for the interaction of atomic shells provide the current "d" shell spherical harmonic modeshapes for various spatial axes and are shown below.

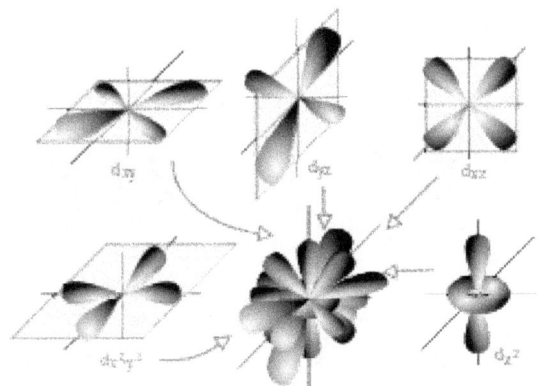

Some of the above graphical images show relatively smooth individual spherical harmonic modeshapes individually comprising an atomic surface for a particular harmonic shell configuration. Below are computer-generated figures of a specific spherical harmonic. It is these types of modeshapes of specific harmonics which give an average curvature to a total sum atomic spatial curvature. Hence the spacetime curvature will be more pronounced for element unique properties and characteristics. However, as we will see in the "electron" structure, some spacetime harmonics will independently resonate with very specific spacetime harmonics across the atomic boundary with other atoms, known as covalent, where atom spatial boundaries are in resonance and the frequency-time permit the individual spherical harmonic lobes to complete the symmetry of incomplete harmonic lobe harmonic frequency "subshells" in neighboring resonant atoms, at times forming what is currently called "covalent" bonds.

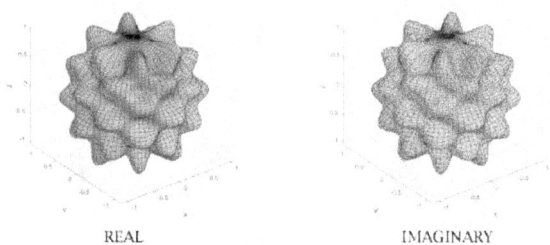

Figure 1: An illustration of the (left) real and (right) imaginary components of the spherical harmonic $Y_{12}^{6}(\theta, \varphi)$.

Repeating here, the natural frequencies of a sphere can be found from the Laplace Equation. The real angular portion, Laplace spherical harmonics Y_{ℓ}^{m}

can be visualized by their zonal, l, and sectoral, m, indexes in the figure[28] below as:

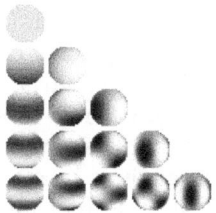

for $\ell = 0$ to 4 (top to bottom) and m = 0 to 4 (left to right). The negative order harmonics Y_ℓ^{-m} are rotated about the z-axis by/with respect to the positive order ones.

    To re-iterate an important feature, the modeshape curved boundaries are space and time surfaces, they produce forces that are also dynamic. The total of the modeshapes curvature, ie Fourier sum, is proportional to the chemical property determinative when not in precise resonance with neighbors since the average boundary can be used. However, the individual spherical lobe surface modeshape curvature is determinative when atoms resonate in the same lobe frequency or harmonic, wherein the curvature of the resonating modeshapes predominate in the attraction or gravity. This determines the type of bonding, the difference between ionic and covalent, hybrid, and others that atomic neighbors will exhibit.

---

[28]   Wikipedia "Laplace's Spherical Harmonic"

What's the charge?

No treatment on an atomic structure can be complete without addressing charge. But **"No one knows** what electric charge is or why there are two kinds of charge, positive and negative as they are commonly called.."[29] A charge is a property that is well used in many models but its physical structure or property is not understood but only represented by label and convention. In our physical model, charge occurs where spacetime curvature has 4D asymmetric or eccentric body surfaces of different spacetime curvature in their boundary, atoms, subatomics, and molecules, have fields from spacetime curvature warpage attraction. Eccentricity vector processes its motions as well, giving osculating eccentricity vector continuous change. In addition, 4D symmetry impedes the general spacetime curvature attraction with uniformity of curvature to all body surfaces. This is caused by the uniformity or curvature, 3D symmetry, which attracts equally from any surface point, so any rotation, vibration, or translation will prevent gravitational adherence to any particular surface point. The Nobel or inert elements, along with neutrons, neutral pi mesons, neutrinos, and the like, are the most symmetric-3-spheres, and thus "neutral". They cannot form charge because their atomic surfaces, spacetime curvature boundaries, have 4D symmetry about a center, complete rotational symmetry, and hence have no point on their attracting surface whose local curvature is greater or less than any other point on their atomic surface. Therefore they are perfectly content to stay single, rolling, spinning, and rotating

---

[29]   H.M.Schey, *Div, Grad, Curl, and all That*, pg 5

out of any proximate attracting surface curvature. We perceive that as not "reacting" with any other particle or staying "inert" or "stable".

Electric charge is an attractive field that we measure and have assigned all particles with numerical measured values of such a property, yet physically it is an asymmetry in the particle boundary assigned a value to certain subatomic particles, +1 to a proton, positive curvature entity, and -1 to an electron, mostly negative curvature entity. Since these have the same or similar spacetime curvature boundaries, these arbitrary assignments enable physicists to provide a normalized spectrum of charge, to a reference of zero.

But atoms also have different "charge magnitudes" and it isn't as simple as a plus or minus integer one in valence. Inert or Noble elements have a neutral charge or zero valences. This is traditionally explained by the full outer orbit theory. Since we now know that electrons are not orbiting a nucleus, how does a vibrating alternate spacetime sphere model explain the valence and charge of the periodic chart-organized elements?

We repeat here because it is so important, that charge is a geometric property determined by spacetime boundaries comprising spacetime asymmetric curvature about a particle center or geometric object center. Thus charge is described as a spacetime curvature boundary of relative positive or relative negative curvature producing an attractive or by convention repulsive force from neighboring relative curvatures. A charge is not mapped from negative curvature space to negative charge or positive curvature space to positive charge. Most charge occurs, relativity/attraction, in negative curvature space. But since the curvature of spacetime is relative to for example other negative curvature space, some curvature will present a positive charge to a relative "zero" reference, while other curvature will present a negative charge, determined by the convention of naming the zero and direction of flow.

As stated above, the curvature of atomic and subatomic boundaries are dynamic, undulating 4D membranes between disparate spacetimes. Thus the curvature of atomic boundary varies in accordance to their component harmonics, and an atom's specific electrical charge will be determined by the atom's harmonics, "electrons", which translate to physical modeshape boundaries. Atom boundaries with asymmetric resonant harmonics will exhibit perceived curvatures of these specific resonant harmonic boundary modeshape curvatures to other atoms with asymmetric but resonant spherical harmonic modeshapes. In the figure below the modeshape curvatures are established by the dominant harmonic radii in resonance with other atom dynamic modeshapes, defining the curvature and relative attraction between them.

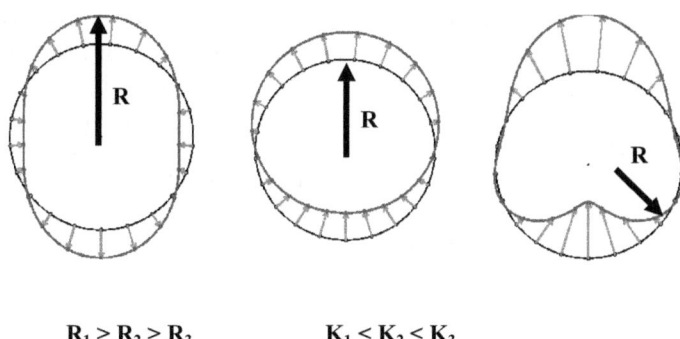

$R_1 > R_2 > R_3$ $\quad\quad K_1 < K_2 < K_3$

As it stands by virtue of curvature magnitude, the charge from the shape with the smallest radius of curvature would be largest because the curvature is greatest. Moreover, since the middle modeshape radius lies between R1 and R3, there will be a middle magnitude attraction from all sides.

This also describes some spatial aspects of the nature of certain types of "bonds" and atomic bonding. The more precise mechanism of bonds will depend on the extent of the symmetry of the spherical harmonic's modeshapes, at their harmonic frequencies, or time, which will drive the overall curvature of the atomic boundary to neutral, inert elements, as the harmonic's modeshapes will attract to all sphere points with equal preference and the translating-rotating motion of the atoms will be sufficient to preclude any "bonding," but only inelastic collisions will occur where there is some attraction.

Thus there are at least two temporal dependences to charge and bonding. One temporal bond type is the macro time is averaged over all of the atom's harmonic modeshapes for total atomic boundary curvature. The total boundary is integrated over the discrete harmonic boundary curvature. The other bond type is the responding modeshapes dominating at certain intervals, producing "covalent" or harmonic resonance bonding, where the co-operating atoms have the same attractive curvature at the same time, lock-on.

Valence or Oxidation State

In the Chronature model, valence is all about symmetry and the degree to which the sphere geometry will approach or depart from a 3-sphere's symmetry. Elements are listed in the Periodic Table with their characteristic Oxidation State demarking the number of "outer shell electron" states available to be "filled" or "borrowed". What is physically happening here, since electrons are not orbitals? Since electrons are atom membrane boundary harmonics, solutions to the vibrating sphere equation, we know from simple mechanics that dynamic motion from an external forcing function can only be imparted to abounding structure in the natural frequencies of that structure. The atom's harmonics come from the proton nucleon movement within. The atom is a structure with natural frequency harmonics fitting the solutions to a vibrating sphere. Electron shells traditionally have been given numerical designations for sets of "electrons" called s, p, d, f, historically for the early studies of atomic spectral lines called Sharp, Principle, Diffuse and Fundamental.

The oscillating flow of spacetime curvature impinging on a one-sided atomic membrane boundary will cause the impinging oscillatory frequency content in the frequency region of any available natural harmonics to be excited, transferring the curvature in frequency between the two systems. But where the forcing harmonic oscillator comes from within the atom, from a "proton" particle oscillating inside from wall to wall, the atomic boundary will have just one harmonic for each harmonic oscillator, proton. But from a standpoint of vibrating sphere harmonics, where the "valence" or electron orbital has one "available electron orbital", only one impinging compatible frequency harmonic can be added for sharing in resonance between another atom with compatible open harmonics. The impinging spacetime oscillatory curvature is then transferred to the atom, not "filling its outer shell" but adding-subtracting the spherical harmonic in the set which is currently not excited, hence "available." Hence the atom's natural harmonic structure gets filled from the lowest frequency to the higher frequency in accordance with impinging curvature to be transformed to resonances available in unfilled natural frequencies; only those, providing atomic boundary change. Thus the model will appear more like the depiction below, where a "squashed" down Oxygen will form the center of the "molecular" system with "orbiting" or resonating Hydrogen. An "orbit" is a geodesic curve surrounding the Oxygen in which the Hydrogen experiences no forces, and hence no change in momentum and angular momentum remains constant. Unlike the celestial mechanic model, atoms are free to collide, from attractive spacetime curvature surface fields and forces, with each other and be pushed apart from frequency and or frequency-time dissimilarities. This excitation is generally what we call "heat" and is the molecular movement that changes the surrounding spacetime such that the momentum is no longer constant. These time dissimilarities act like something from one time attempting to interact with something from another time, their curvatures would be disjoint and disabled. However, these all work in concert to impart a superposition contribution of push-pull between the atoms causing the "bond".

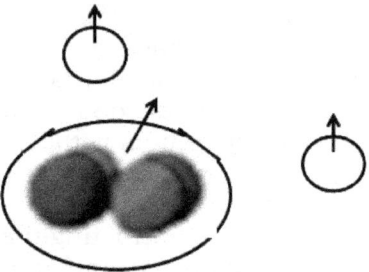

The arrows pointing up reflect that spin about a center does not encounter any forces or momentum since the motion is geodesic in spacetime. However, the spin does carry angular momentum and can interact with other particles. Sometimes this spinning hydrogen is called a polaron.

What is the current theory on atomic bonds from electron orbital mechanics?

In addition to his Exclusion Principle, Pauli's theory that no two electrons in an atom can have the same quantum number, Pauling is also the author of Valence Bond Theory and Hybridization, the first plausible model applying QM to the Bohr model in explaining where electrons can be found orbiting the nucleus during atomic and molecular interactions. The Valence Bond Theory, states that a covalent bond, that is where the atoms share an electron more-or-less equally, developed when two half-filled electron orbitals "overlap" containing two electrons of opposite sign. This "shared electron pair" is most likely found, remember that this "orbital" is the volume or probability "cloud" where an electron can be most of the time, is in the space between the two of the atoms forming the bond. Why the "electrons" in this shared volume don't collide and collapse both atoms is unknown. This should tell us that this current model, a breakthrough that it was for the time, was never complete or even understandable. The picture below shows how this would appear in physical reality if it were a Carbon and 4 Hydrogen in what's called sp3 Hybridisation.

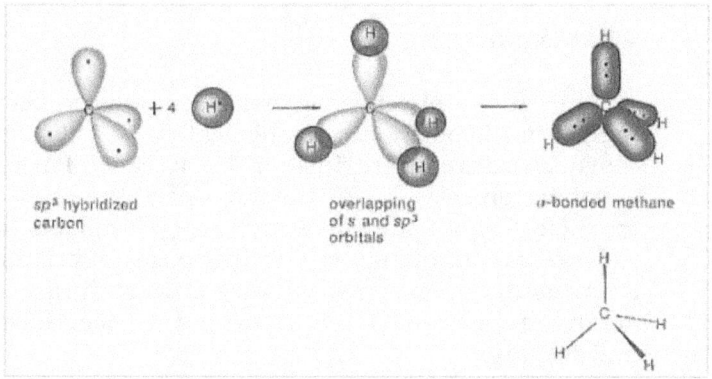

According to the current general QM models of the atom, the interacting "electrons" are not shown, only the volumes that they "may be found in" by probability are shown. In the Chronature model, the s, p, d, f "shells" are modeshapes derived from a vibrating sphere model harmonic solutions and the modeshape solutions are enclosed surfaces which are the "electrons", not probability volumes bounded by the lobes or leaf probability structures. Spoiler alert, the surface curvature is the spacetime curvature that directly causes the acceleration forces we measure as electro-negativity or bond strength. Chronature electron interactions will show the surface's interactions by proximity to surface to surface, as the spacetime curvature will be attractive to the point of surface touching and repelling, Exclusion Principle, upon entry. This will set up a vibration between the atoms at a particular characteristic frequency and creates molecular vibration modes which are measured today by IR radiation and other methods.

What is parity?

From a purely geometrical matter perspective then, what is parity in the physical world? All "particles" have an intrinsic property called Parity, which is "preserved," for the most part. We say "particles" about the currently accepted models. In the interests of brevity and for purposes of building on measured basics, we will now proceed further about what is known about parity, the exceptions, and "violations."

First a short definition of parity from Wikipedia,

"In physics, a parity transformation is the flip in the sign of *one* spatial coordinate. In three dimensions, it is also commonly described by the simultaneous flip in the sign of all three spatial coordinates:

$$P : \begin{pmatrix} x \\ y \\ z \end{pmatrix} \mapsto \begin{pmatrix} -x \\ -y \\ -z \end{pmatrix}.$$

A 3×3 matrix representation of a point **P** would have a determinant equal to −1 and hence cannot reduce to a rotation that has a determinant equal to 1. The corresponding mathematical notion is that of a point reflection. Adding spacetime coordinates gives us yet other possible reflection combinations. The t and −t, ω and −ω, flipping signs are missing but represent the physical characteristics of parity changes in rotating spacetime entities and interactions.

In a two-dimensional plane, parity is not a simultaneous flip of all coordinates, which would be the same as a rotation by 180 degrees. The determinant of the P matrix must be −1, which does not happen for 180-degree rotation in 2-D where a parity transformation flips the sign of *either* x or y, not *both*.

The spherical harmonics have well-defined parity in the sense that they are either even or odd with respect to reflection about the origin. Reflection about the origin is represented by the operator

$$P\Psi(\vec{r}) = \Psi(-\vec{r}).$$

For the spherical angles, {θ,φ} this corresponds to the replacement {π − θ,π + φ}. The associated Legendre polynomials $(-1)^{\ell-m}$ and from the exponential function we have $(-1)^m$, giving together for the spherical harmonics a parity of $(-1)^\ell$:

$$Y_\ell^m(\theta,\phi) \to Y_\ell^m(\pi - \theta, \pi + \phi) = (-1)^\ell Y_\ell^m(\theta, \phi)$$

This remains true for spherical harmonics in higher dimensions: applying a point reflection to a spherical harmonic of degree ℓ changes the sign by a factor of $(-1)^\ell$."

In another aspect of parity in physical representation, the figure below depicts the space spin axis as the solid arrow and the frequency-time axis as the dotted arrow. Although only for combinations or configurations are shown one can see that this two orthogonal axis can have many unique combinations operating on a sphere together. Our guess for the physical reality is that parity of one is reached when space and the time spin vectors are aligned and negative when the spin vectors are opposite of off-angle.

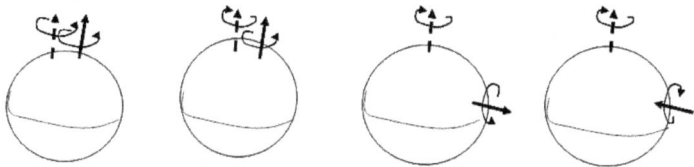

The big deal is parity is generally conserved, but not in "weak" interactions between elementary particles. We won't go into left or right-handed systems, but they seem to make a difference in weak force interactions. Some kind of geometrical spacetime curvature symmetry property appears to us as "parity". Looking at frequency-time and the negative, we can get conservation of spacetime curvature or not depending on the spacetime curvature configurations involved in the interaction. But again here, it is not just spatial symmetry.

Spin

It's been known for a long time that even at rest electrons were found to have a magnetic field, which would classically mean that the electrons are spinning balls of charges that is how they produce a magnetic field. Under the Chonature atomic model, the magnetic field comes from an atom having 2 opposite but equal spacetime curvatures that serve as opposite poles, an attracting and a repelling, for a magnetic field. In QM the electrons are not spinning balls, they have an intrinsic magnetic field which is not because of any motion. Interesting how that complies with the Chronature model. So today's physics parlance, the magnetic dipole is called spin, and the value is based on the "charges and the amount of magnetic field created". In the Chronature model, spin is therefore the strength of the atom's magnetic field created.

Much more can be found here, but we can see that parity has to do with symmetry of object harmonics and objects symmetry in its harmonic reflections and rotations wrt to space and time, not just space. This affects changes and change states and is particularly useful for spheres transforming. Why spheres hold such large importance will become apparent. Knowing parity and symmetry properties of an atom or subatomic particle, which are measurable, reveal much about their configuration and geometrically dictate some very important properties of attraction and bonding.

WRT Chronature, most of the parity is the degree of physical alignment between the outside atom spatial rotational axis and the atom inside volume temporal frequency-time spin axis.

What's the deal with Spin and what is it physically?

The Standard Model divides all particles into two camps by "spin", Bosons and Fermions. The Bosons are the force carriers called Higgs, gluons, photons, W and Z. These all have "integer spin", 1, 2 ... Quarks, electrons, and neutrinos are Fermions, they have a "half unit of spin." Somehow these clues all come out of particle collision data and QM. But one thing is certain, currently spin has no mapping with physical reality, but it is represented to appear as if it does. The current explanation and use of "spin" come to us from QM and is a quantum of angular momentum of a particle, yet another non-physical quantity because infinitely small point-like particles cannot have measurable physical spin. QM puts yet another misleading analogy into a name to make one forget that this model has little to do with physical reality. We say this here again to help us keep our focus on a model of physical reality, not probabilities and states.

In the Chronature model, spin is a very physical quantity, the vector of the physical spacetime rotation of the atom space within and relative to the rotation of the spacetime boundary enclosing that atomic spacetime within. Because frequency-time inside the atom manifests in a bounded spacetime as cyclical as well as spatial rotation, there will exist a spin axis about which this inside the boundary space manifestly rotates somewhat independent of the temporal cycling. This boundary spin axis will not be the same as the rotation of the spacetime boundary, skin, because that spacetime boundary has a discrete and different curvature which complies with the macro spacetime in which it resides. Hence one can have parity of 1, temporal and spatial axes alignment, -1, 180 degrees, or +- ½, +- 90 degrees between the rotation axis vectors. Because spherical harmonics carry symmetry about axes, spin axis and hence spin vectors will have discrete right-angle relationships of +- 90, +- 180, +-360, etc, degrees.

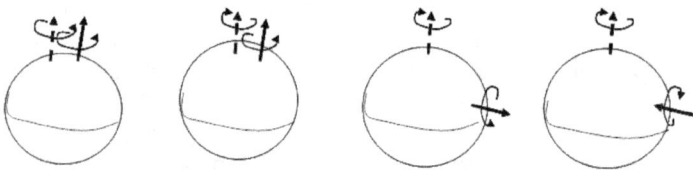

Since boundaries are spacetime curvature, boundary harmonics can independently establish physical forces depending on any temporal parameters which give rise to resonance between interacting surface entities. Most of these are currently modeled through "angular momenta" from the old Bohr model colored by QM.

In addition, to build on the current usage of the "spin" mechanism spin-spin interactions are used. Classic physics defined, spin-spin coupling as the coupling of the intrinsic angular momentum, "spin", of different particles. Such coupling between pairs of nuclear spins is an important feature of nuclear magnetic resonance spectroscopy as it can provide a type of information about the structure and conformation of molecules. Spin-spin coupling between nuclear spin and electronic spin is responsible for what is known as Hyperfine Structure in atomic spectra and a well-known physical parameter. In addition, the orbit and spin of a single particle can interact through spin-orbit interaction, in which case the complete physical picture must include spin-orbit coupling. Also, two charged particles, each with well-defined angular momentum, may interact by the atom boundary harmonic attractive forces, in which case coupling of the two one-particle angular momenta to a total angular momentum. In both cases, the separate angular momenta are no longer constants of motion, but the sum of the two angular momenta may still be useful in modeling the motion.

By physical analogy, in astronomy, spin-orbit coupling reflects the general law of conservation of angular momentum, which holds for celestial, hence physical, systems. In simple cases, the direction of the angular momentum vector is neglected, and the spin-orbit coupling is the ratio between the frequencies with which a planet or other celestial body spins about its axis to that with which it orbits another body. This is roughly shown below as "planets", Hydrogen, about a sun, Oxygen with the "spin" and the "orbit" paths depicted by s and o arrows respectively. The angle and distance shown between the atoms are of course averages and only gross averages measured with the physical assumptions and constraints of the present static measurement apparatus. Also the 3rd and 4th dimensions are not shown but the angular momentum vectors would be out of the plane.

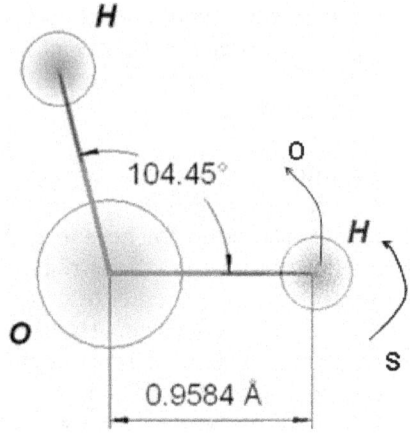

In the present physics nomenclature, this is more commonly known as orbital resonance. Analogous to heavenly bodies with proximate spacetime distortion causing gravity, the underlying physical effects are tidal forces. By analogy but with much greater force strengths of interaction, orbit and spin of a single atom or more, the attractive qualities will manifest in various spacetime dynamics, some of these can be modeled and measured by total angular momentum from spin-orbit mechanics from boundary harmonic specific modeshape curvature and the coupled rotation dynamics occurring throughout the interaction.

What does all this spin-spin and spin-orbit mean for atomic spacetime entities'? In atomic elements with few boundary harmonics where mostly the nuclear entity forcing function frequencies are dominant, the atom boundary harmonics can be treated independently of each other; mathematically their spin "operators" are conserved.

How can we use the constants and parameters for resonance calculated substances currently? In the current model, the operating, Larmor, frequency ω_0 of a magnet is calculated from the Larmor equation:

$$\omega_0 = \gamma B_0$$

where B_0 is the actual strength of the magnet in units like Teslas or Gauss, and γ is the gyromagnetic ratio of the nucleus being tested. Chemical shift δ is usually expressed in parts per million (ppm) by frequency because it is calculated from the difference between a resonance frequency and that of a reference substance divided by the operating frequency of the measuring spectrometer. All of these measured parameters have real physical manifestations which are combined with the harmonics of each atom to create resonances that define each "bond" with its unique characteristics taking into account the spins, spin axis, orbits, individual atomic harmonic contribution to spin and orbit.

Analogous to satellites about planetary bodies, atoms will eventually park themselves in what's known as "geostationary orbit", a neutral position away from the attracting body. There, it will stay "locked" into the "atom site" ie. the "planet's", rotation. This of course is due to the superposition of the space-space geometry and the harmonic frequencies between the atoms. The rigidity or fluidity of these interactions is the manifestations of state; solid, liquid, gas, plasma, or condensed matter.

What models are currently used for calculating molecular mechanics ?

FIR Infrared Spectroscopy – Signal, Shape, Intensity and Functional Groups are the short answer.

Bond Vibrations, Infrared Spectroscopy, and the "Ball and Spring" Model are predominantly used today. Typical results can be seen below for water vapor

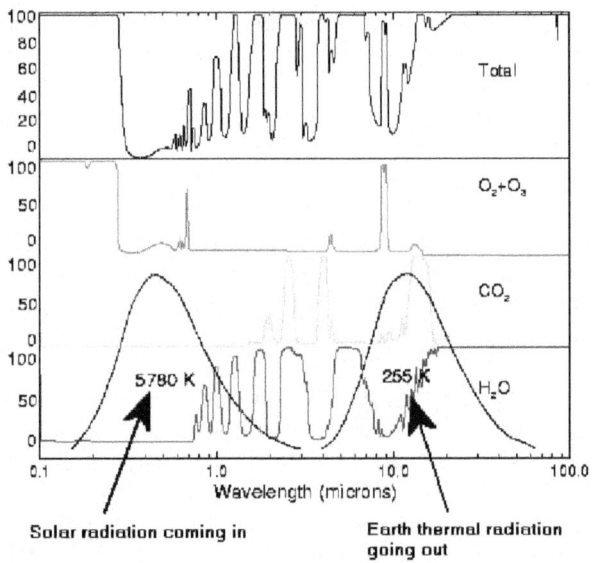

The wavelength for the various atomic components is systematically discovered by the IR spectrometer and is indicative of the type of atomic interaction present.

The interacting atoms in their molecules can exhibit vibration, stretching or bending mechanics, or rotational and present discreet properties of absorptions or transmission under certain IR stimulation, at particular wavelengths.

For us, the Chronature model of spacetime curvature attraction between atoms holds the atoms together in their particular "bonds", which are nothing more than local spacetime telling the atoms how to move, while the atoms are telling the local spacetime how to curve. Using Chronature models these can be predictable for all kinds of conditions not just measured under very narrow ones to ascertain bonds, optical pumping, band-gaps, QCL, LEDs and the other phenomena.

What is the atom's skin or surface boundary made of and what are its properties?

Mathematically the atom's boundary is an infinitely thin 2D surface in 4D spacetime. But physically, the spherical wave equation solutions tell us that the atom's boundary or skin has density, thickness and is continuous, having spacetime curvature perturbations from within and without moving it in and out from the atom sphere center, about where the nucleons wander. Also, we know now that the skin density is the degree of curvature of the spacetime at the border of two discrete and different spacetimes, with opposing curvatures, hence the change properties would be discretely different, analogous to a shock between fluids. Moreover, we can see from photographs of molecules that the atom boundary is very flexible, deformable, and manifest a fairly well-defined thickness boundary. Of course, some of this is due to the microscopy technology but never-the-less the reality is there is some thickness and it is finite.

As it turns out, atom boundary while a vibrating sphere, is sufficient to derive the atomic electron shell structure, it is insufficient to explain other properties of atoms, such as magnetism and dipole moments. Returning to physical reality once again, we introduce a spacetime bounded Froloff Sphere or F-sphere, after its inventor, essentially a Klein bottle morphed into a spherical configuration with a two ply-sphere whose inside and outside surface are actually and physically the same side of a one-sided surface, providing both a negative curvature and a positive curvature on the same side of the same one-sided surface boundary producing a dipole structure. This would provide magnetic properties with a logical physical atomic spin, as the poles could be physically where the atom boundary flips to make a one-sided surface at top and bottom, see figure below, nucleons not shown. All while having the spherical harmonics the provide all of the electrical, chemical properties of an element atom.

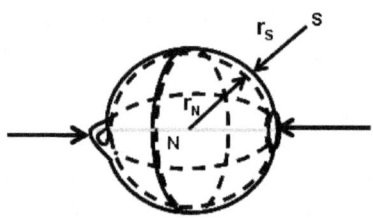

F-Sphere

Please note that this F-Sphere sphere configuration shows a relatively small handle or umbilical at the F-sphere middle shifted bottom. This handle virtually disappears by morphing into a tiny dimensional twisting appendage to create a spheroid essentially a one-sided sphere with a North-South pole spin axis in some orientations in spacetime. The throat section shown at the top is morphed to a point or smaller for a north pole with the usual Klein bottle Mobius transformations. While this may cause heartburn to the mathematicians, physicists will never-the-less love it for its physical explanation of reality from geometry alone. Moreover, these little one-sided surface twists and turns become more important in explaining physical spin harmonic states and bonding. The "north" and "south" poles also give rise to longer and shorter lines of spacetime which then give rise to negative and positive curvature, for attractive and repulsive forces, on the same side of a one-sided surface. This as shown above for a magnet, and occurs as the geometrically straight lines progress from shorter to longer lines, south pole, from negative curvature, and shorter lines at the "north" pole, positive curvature, giving rise to the attractive and repulsive forces in spinning atomic entities respectively. Magnetic dipole moments and magnets arise from these spacetime curvature configurations and their combinations and orientations which are additive to what is currently called "magnetic domains."

Topology gives us a leg up on understanding what is physically happening when atoms collide and their surface boundaries deform, to the point of breach. Where either atomic boundaries or "bonds" are broken. Where a submanifold is "closed interval" and of finite length, where two conditions are met, I would conjecture the atom 3-spheres are topological properties considered "compact". From that the mathematics of topology every continuous image of a compact space is compact; and that every compact finite-dimensional manifold can be embedded in some Euclidean space. Furthermore, that the product of compact spaces is itself compact. Perhaps collisions can be modeled as quaternion products from a "distance" but deformable membranes upon "contact". This would indicate that colliding atoms, compact spacetimes, are compact and homeomorphic. This would indicate something about the atomic boundary deformation or morphing that occurs during atomic collisions. Deformation of the atomic sphere boundary occurs and calculations of curvature can include the "elasticity" of the atom boundaries due to topological deformation from neighboring or pairing atoms. For example, the Hydrogen atomic radius, un-deformed, is measured as 5.1 Angstroms but the diatomic Hydrogen is measured at 3.7 Angstroms, due to the radial shrinkage from the "covalent" bond or squishing the individual Hydrogen atom boundaries together at a resonant modeshape harmonic, ie since all Hydrogen has the $1S^1$ harmonic, they would naturally resonate with other Hydrogen atoms at this modeshape to complete the 1S harmonic set.

     Also, topology metrization theorems give the necessary and sufficient conditions for a topology to model and describe the actual physical interactivity, something needed to model the spacetime surface dynamics correctly in the metric tensor.

     The twisted one-sided atomic 3-sphere boundary likely has other topological properties such as diffeomorphism, isomorphism, and more morphisms than you can shake a nanotube at, depending on surroundings conditions. In addition, the F-Sphere has a hole at the north pole which throws it into the Euler Number 0 group of Topological structures. These may play a larger role in EM, where the F-spheres become chained, translating frequency-time to the macro spacetime monotonic time dimension producing stretched-out F-sphere car train. Another way to say that is the atomic structure F-Sphere is Topologically Isomorphic with EM spacetime.

Is not a Klein bottle of the topological non-orientable class and a sphere of the orientable class object and does this mathematical topological distinction violate the morphing rules of object shapes from one shape to the other?

It is precisely the non-orientable property of the Klein bottle topology that changes the normal vector to opposite directions upon surface traversal. So that is where we need start to get two equal and opposite poles on the same surface to form magnetic fields. So we configure to a sphere that has some adjusting property that gives it the Euler property of Klein bottle. We are talking about something called Homotopy, where objects are homotopic if they can be deformed from one to the other, without cutting or tearing, where one can be continuously transformed into the other. Homotopy can be mathematically and topologically very tricky. For transformation from a torus to a sphere, for example, shows that there are certain cases where a sphere can be obtained from a torus. This is the famous Alexander horned sphere in 3D Euclidean surfaces are constructed with a standard torus, whereby the Alexander horned sphere is also homotopic to the standard 2D sphere. Stitching manifolds together, using the Thurston Geometrization Theorem, allows the topological gymnastics to obtain shapes not thought possible before. Moreover, Perlman in 2003 proved the conjecture using something called Ricci flows to determine whether various geometries were equivalent. Showing that in local areas 3D surfaces such as spheres and torus's can be stitched together from finite closed manifolds using Euclidean patches with attached loops. Furthermore, connecting all these high fluting mathematics to physical reality, Ricci flows are also instrumental in the derivation and application of the EFE.

    But to conclude here, the Klein bottle is a closed surface, non-orientable, mathematically compact geometrically speaking and has no boundary. Hence a normal vector traversing the one-sided surface will flip on a round trip, producing two equal and opposite poles, spacetime curvature attractive and repulsive. Its Euler Characteristic, a group classification artifice, is zero like in a Torus. On the other hand a sphere also has no boundary but is orientable, compact, has a surface with two sides, and not capable of sustaining a physical dipole moment in a spacetime curvature configuration as a sphere. If you believe that the mathematics must allow and predict what we find in nature, then it follows that a sphere is topologically transformable to a Klein bottle dipole object and vice versa.

What about some pictures of atoms, with all-out high-powered new and emerging microscopy technology, don't we have a picture of an atom yet?

A team from IBM in Zurich using a variant of a technique called Atomic Force Microscopy (AFM) first published single-molecule images of a molecule called pentacene, so detailed that the type of atomic bonds between their atoms can be discerned. Please note that no "electron clouds" are visibly interacting in the bonds. The thickness of the boundary between the atomic spacetimes appears discrete and finite.

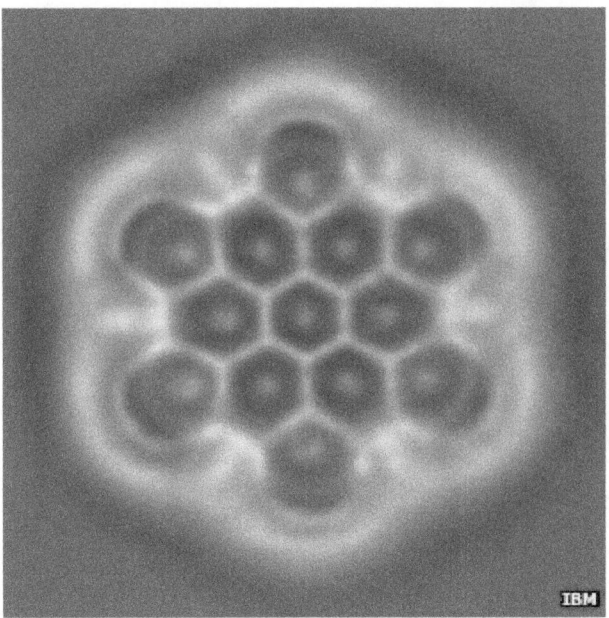

The boundary "bonds" at the molecule's center appear shorter, thinner lines, crisper than those at the edges. We speculate this is because they involve more neighbors squeezing in from more sides. However, the tightness of the symmetry towards the center can perhaps explain the distance and boundary thickness as well. Note that QM predicts that an observer will capture, perceive, one state. Furthermore, we should see a point for the electron but don't. This should worry QMers; no point or particle electron is to be found... hmmmm

Atomic Force Microscopy (AFM) technology has most recently given researchers in China a view of a hydrogen bond shown below.

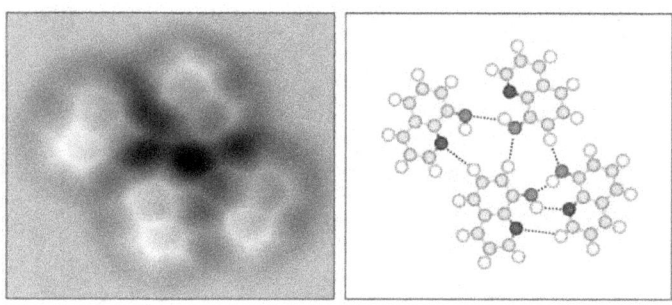

Hydrogen bonds linking up quinolines have been seen for the first time © Science/AAAS

Xiaohui Qiu and colleagues at the National Centre for Nanoscience and Technology, China, used AFM to view 8-hydroxyquinoline (8hq) because it's a flat molecule but a single hydrogen bond would be out of the plane of the structure. Furthermore, their AFM, Atomic Force Microscopy, observations of hydrogen bonds formed between 8hq molecules were not expected, because of the very low electron density in the proximity of these weak bonds. Qiu admits that the nature of a hydrogen bond is still debated. It has long been considered an electrostatic interaction, but more recent models suggest that it has chemical bonding characteristics as evidenced by x-ray diffraction experiments.

AFM does not image atoms and molecules with visible light. Briefly, AFM involves dragging an atomically sharp probe on a cantilever over a surface and measuring the deflection. For example, a non-contact AFM is dragged with a single CO molecule deposited on the end of the tip. In the constant height mode used, the AFM tip is oscillated at the resonant frequency of the cantilever and scanned across the surface at a constant height above the surface (not touching). Depending on whether the surface exerts a force (electrostatic, van der Waals, etc.) on the tip, the resonant frequency changes, producing the signal that is measured. So what we are seeing in the picture is a reflection of the local "electron density." AFM is akin to dragging a finger across a rough surface and feeling the roughness.

As we get closer to view the structure of physical atomic bonds, we get farther away from the accepted model of the Bohr-QM atom model. Pictures of molecular bonds will continue to surprise us because our present atomic model has long outlived its usefulness. A Bohr model calculation of an orbiting Hydrogen electron yields an electron orbit speed of one hundredth, 1/100, that of the speed of light, C. Thus one would expect that the AFM using laser would scan objects at 1/100th their speed as QM collapsed states, not grey or colored shell structures. But physical bulbous shell structures in the pictures are pretty much what would be expected from a Chronature model.

Why do atoms become smaller progressively with higher masses?

The Chronature model needs nothing more than a sphere's harmonics and geometry of space curvature for the physical explanation of what we see in size properties. The chart below shows the elements and their corresponding atomic radii. There is a pattern in the progression of atomic radius vs. element number on the periodic chart. But why does the atom size increase at first, only with progressively higher S shells, but decrease the radius with progressively higher p, d, and f shell harmonics? Remember as in the Fourier Series, the instantaneous locus is a sum of the curves natural sine and cosine frequency component harmonics, which sum to provide the actual curve shape at any point in time and space to define the curve. The flat 2D squarewave function below shows a Fourier construction of the square wave, with only the sum of the first 3 harmonics. Adding increasing harmonics approach the actual square wave.

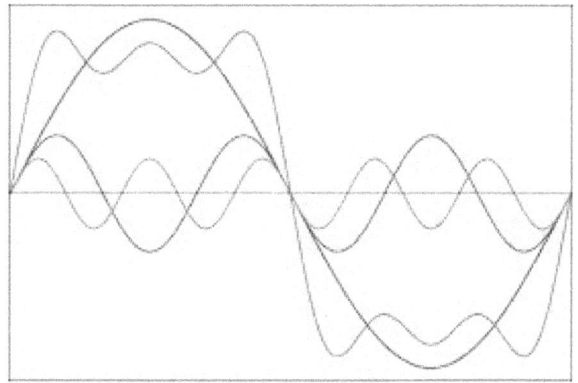

      The answer lies in the difference in spherical harmonic modeshape between the fundamental, S, and the addition of all of the p, d, and f harmonic modeshapes progressively down the periodic chart rows. A Fourier sum would add up to give the total boundary shape and at the higher harmonic additions, converging to a smoother but tighter dimpled sphere shape with more "electrons" or higher atomic number elements. The p, d, and f modeshapes all radial add or subtract from the atom's sphere surface shape. The S modeshape increase at all angles, and hence its contribution would be the largest overall angles, not just smeared over a smaller boundary area of a tesseral, p, d, or f harmonic.

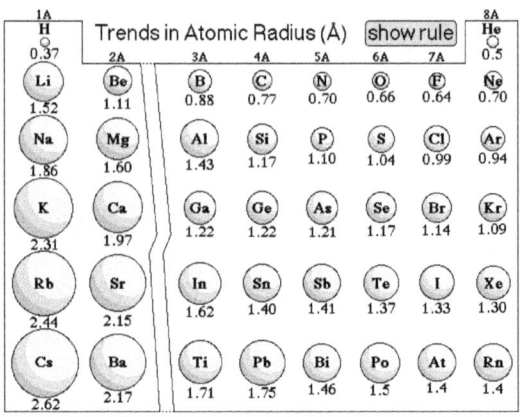

Even the second S harmonic, nS^2, will reduce the average atom boundary radius as it extends inward toward the center of the atom at all angles, subtracting from the first nS^1 S subshell modeshape. However, jumping to the next S shell harmonic does add uniformly to the whole shell positively. Progressing higher harmonics, p, d, f, will spread the atom boundary producing a retreating average shell radius distributed over the entire spherical boundary. This resulting in a smoother atom boundary with more but smaller "bumps" and "dimples" but an overall smaller average radius.

This progression on element radius is only a part of the atom boundary shape, as the radius of curvature at any time and any point on the boundary are also significant factors in the spacetime curvature of the boundary. Spherical harmonic frequencies producing sharp boundary contributions will produce asymmetries that will make curvature more attractive at those positions.

The electron "orbital" radii are the atom spacetime physical boundary as snapshots in time, but when these harmonic mode shapes are Fourier summed they give the average atom radius. Fourier addition per orbital in an element (harmonic modeshape) provides the average atom boundary radius. eg

H : $1S^1$
He: $1S^1 + 1S^2$
Li: $1S^1 + 1S^2 + 2S^1$
Be: $1S^1 + 1S^2 + 2S^1 + 2S^2$
Bo: $1S^1 + 1S^2 + 2S^1 + 2S^2 + 1P^1 + 1P^2 + 1P^3$
:
:
Ne: $+1S^2 + 2S^1 + 2S^2 + 1P^1 + 1P^2 + 1P^3 + 1P^4 + 1P^{5+} 1P^6$
etc as per periodic table

The basic wave equation coefficient $1/c^2$ is:

wall_density/wall_surface_stress =
$$\text{permissivity} * \text{permittivity}$$

The wave equation alpha constant for the basic spherical wave equation is the classic constants that we know and have are functions of permissivity and permittivity.
.

The classic explanation is that the stronger attraction from the stronger charged, more proton nucleus, brings the "electron orbits" closer to the atom nucleus, thus reducing the atom's overall electron orbiting radius. But contrary to this, all bets are off when, the progressive $1S^1$ shell electrons appear in the 1A column element, as the radius grows again. Not only that, but classically we still don't know what "charge" represents physically but use circular logic in using it to define its effects.

4 – Electrons and Electron Organization

J.J. Thompson won the Nobel Prize in Physics in 1906, when he showed that electrons were particles. His son won the same prize in 1937 for showing that electrons are waves. Like the blind men attempting to describe an elephant, this was another clue that we use a very poor model of the atom and matter in general, as the modelers cannot decide whether what they are "feeling" is a particle or a wave and decide it's both. Whether the electron is in wave state, particle state, or combination electron wave and particle state, the QMT addition to the Bohr model doesn't buy us anything more from a physical structure of matter standpoint. The fact that there are waves in the particles or particles in the waves doesn't increase real physical understanding. Certainly, it doesn't help in the math models either. The truth is there is no electron particle, there never was. It was just a poor model we were stuck with.

With a stark difference to model elusiveness, the vibrating spacetime bounded 3-sphere model gives the same solutions and measured electron shell structure as the current patched up Bohr-QM electron orbital model but without any angular moment assumptions or probabilities. See Menzel reference above, where the rigorous mathematical physics scrutiny is found under the vibrating sphere example in the literature. Inserting the currently known Hydrogen average radius, average, should give us the effective stress density of the atom physical sphere surface boundary and there are other ways whereby we can obtain the physical constant parameters to enable this new spacetime vibrating sphere model useful for numerical processing of actual atomic forces and motions from known surfaces, atomic harmonic modeshapes. Barring that we are stymied by the currently accepted electron shell models having the "electrons" in physically impossible orbits probabilistically partitioned in discrete volumes represented by the spherical harmonic lobes for particular sub-shells s,p,d,f, which fail in providing an understanding of physical reality and eventually application of physical law to gain model understanding.

Current theory has it an electron is neither here nor there but there and here at the same time. In physicist jargon, it's referred to as "non-localized". That's when we are vibrating extremely fast and so we are here and there simultaneously. Never mind we've noticed that atoms still have some actual physical structure, they comprise all things and exist in a vibration physical state. Our limited physical experienced brain can still understand that, no need to abandon logic and physics to pure mathematics on inherently primitive models.

There are several explanations for the electron structure models historically. QM claims that the "electron" is a point "particle" with a probability of being within a harmonic modeshape shell volume ie, the electron is orbiting the nucleus but is most likely found in a spherical harmonic lobe volume called a "cloud" emanating from a nucleus. Model proponents decree that all models are wrong but some are useful. But how these orbiting electrons that must collide with other orbiting electrons do not accelerate and fall into the nucleus or get knocked out of their orbit, maintain stable atoms, is the $64K question, and their usefulness is now a burden, not a help. And why is it that forces affect some matter but not others matter? For example, electrons do not feel the Strong Force yet protons are confined to the nucleus very tightly by the Strong Force.

Chronature holds that the electron orbits do not decay because obits do not exist, or at least as particles, but only as physical sphere harmonics on a vibrating spherical boundary between two spacetimes. In the Chronature model, electrons and electron structure can be explained simply from a Fourier series of vibrating sphere wave composition. The atom electrons are a Fourier series sum of the spatial 3D harmonics of a vibrating sphere, giving the atom its effective boundary with spacetime curvature in space and in time. From Fourier, for a simpler string, 2D harmonics are shown below.

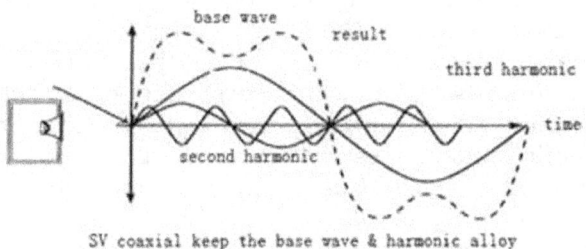

SV coaxial keep the base wave & harmonic alloy

Carrying forward that atom electrons are a Fourier series sum of the spatial 3D harmonics of a vibrating sphere, a rough Fourier for a 3D harmonics modeshapes for the first 10 elements are shownn below.

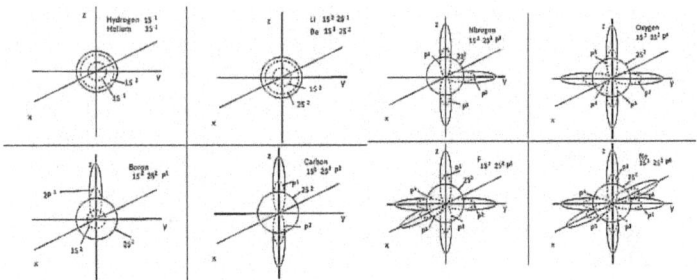

By inspection, we can see that the modeshapes of the harmonics for an element atom have different but related and progressive curvatures for the separate resonance, harmonics as well as for the total Fourier sum harmonic modeshapes. Also noteworthy is that the Nobel elements are symmetrical in spacetime, hence their curvature over the entire body has a canceling effect of attraction, meaning zero. Manipulating the curvatures of these atomic modeshapes in spacetime will provide insight as to what we call Energy, Momentum, and matter continuity, in its many forms. What is not shown is that the amplitudes are calculated from the spherical harmonic frequency associated radius, and must be factored in when making the Fourier sum at each point. The phase is taken into account for these first elements in the series by inspection. That is that the modeshape harmonic is radially symmetric, s shell, or lies on an axis, p shell, or some symmetrical angle, d and f.

Vibration underpins all matter in the universe. Matter does not exist without vibration. The Chronature atomic model analogously depicted here with fluid in a bowl picture below, is the vibration of composite spacetime entities. To illustrate this from a more visible physical perspective, the picture below from Cymascope.com provides an octave of piano note reverberations in a bowl of water. Here, Frequency is analogous to proton-induced modeshapes within an atom. As shown below the classic octave of harmonic note vibrations is analogous to elements modeshapes in the atomic spacetime configurations.

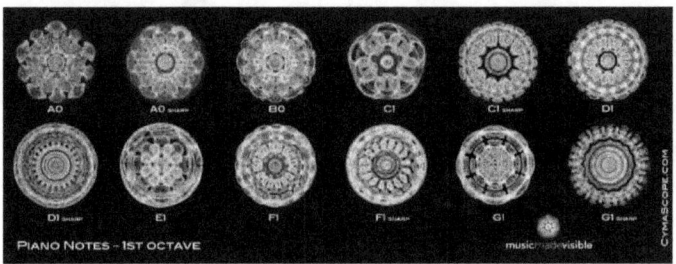

A classic musical note octave is much analogous to a row of elements in the Periodic Table, manifesting modeshape configurations from set composite spacetime entity characteristics and variables.

What does a "free" electron look like?

The above is not the model for a "free electron". A free electron in the Chronature model is a closed surface atomic boundary absent nucleons perturbing its boundary configuration. Without the nucleons, the "free electron" travels lite, providing almost no inertial resistance and hence little or no "mass". The electron spin axis through the poles provides some asymmetry to the free electron, providing it with a curvature that is attractable to some elements depending on their asymmetric curvature nature. Traveling at light speeds outside of a metal matrix or network gives the electron some Lorenze contraction in the direction of travel. Below is a depiction of such a little 'e' beast.

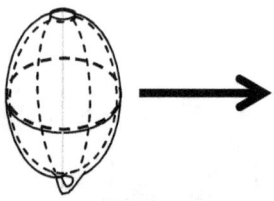

Because the free-electron defines another spacetime, its discrete "spin states" confuses the physicists in making measurements and hence totally confounds and perplexes the classical physicist. However, Quantum Mechanics over time have come up with ways to explain Entanglement Theory, progressing ever more creative in their explanations using all manner of mathematics of tensors and vectors with states, "normalized probability vector", probabilities, conjectures, and more. QMT gets away with this because state transitions happen at light speeds and tiny dimensions, this is virtually immeasurably instantaneous. Note that in reality there is a continuum of transitions through spacetime. But QMT only deals with final states and probabilities of transitioning through these states, time is irrelevant.

Coming back to Chronature, the free electron is an empty F-Sphere and therefore has a 720 ° one-sided elliptical non-Euclidian geometry boundary which then has 8 orthogonal axes, 4 local axes, and 4 macro co-ordinate axes systems. This can be warped in many ways to suit anything we can measure. Hence an electron in a magnetic field will attempt to fall a geodesic straight line, as that is what geodesics are in spacetime. Since there are at least two real-world coordinate systems, the "electrons" local and the surrounding spacetime containing the embedded electron spacetime, the QM model posits "observable" for electrons and positrons is allowed only certain spin vectors of 1, ½ and -1 and the like and without regard to the physical reality of multiple real spacetime co-ordinate systems or the time required to reach these states. Hence QM postulates that there are only those "spin" states and no others in the spacetime continuum and so in physical reality, there is no physical "spin", it's just a dipole that is up or down, North or South. But in the transition of states as measured in physical reality, an "observable", a photon may be shed or positron emitted to balance out the spacetime won or lost in the transition. Therefore spin along an axis becomes a tough concept to understand without "states", but not having access to a physical model prevents the simplification of understanding, forcing QM's to conjure math with "spin vectors", "quantum vectors", "pointers", "eigenstates, "eigenvectors", probabilities and components of each confusing even the most intelligent mind to believe that there exist only "states". The hodgepodge of quantum entanglement math that must be absorbed is enormous, and few make it. When one gives up on their understanding or comprehension of reality for the QM math, some capitulate that the QM model must be correct and their brains are just too small to comprehend this other only probabilistic reality. This is quackery because it forces those without the tolerance for math used as obfuscation, to capitulate and buckle under the pressure of QMT proponents and QMT stultifying limitations. The reader is encouraged to see the fluidic Strouhal fluid-structure interaction analogous reality model for Quantum Entanglement to understand what happens at pair production time with opposite curvature produced in pairs, mostly from symmetry and preservation of spacetime curvature requirements. Hence "spin", or rather opposite direction attributes of phenomena are only a characteristic outcome of the first principle of Conservation of Spacetime Curvature, and free-electron transitions naturally emit discrete pieces of spacetime in the form of photons to maintain an allowed spacetime curvature of the electron at the end state, when forced to align with magnetic produced spacetime. This is also seen as "rotational invariance", where rotating along a geodesic, ie a straight line in curved spacetime, shows transition spacetime curvature is zero.

Using the spacetime of an "electron", how does nature create a magnetic field from spacetime curvature geometry alone?

(a) Starting from a one-sided surface Klein bottle topology

(b) With topological equivalence - shrink the K-bottle handle

(c) Further shrink and migrate the handle to a micro knot at the southern hemisphere and the bottle brim to a northern hemisphere point hole

(d) Note that this configuration only has 2 equal and opposite radii of curvatures and on the same side of a spatial 3D surface, F-sphere.

(e) Shrink the west hemisphere circumference and

(f) Withdraw the inside from the "inside" of a morphed object, this is done in the time dimension

(g) Align the inside and outside sphere curvatures in monotonic increasing time, providing a spacetime curvature object having an attractive and repulsive pole field. QFT has labeled something as "vacuum polarization." This is an observed process in which a background EM field ostensibly produces a "virtual electron-positron pair". Figure part (g) provides a physical model of such an entity, from purely spacetime curvature having an attractive spacetime curvature, electron, and equal but opposite charge repulsive spacetime, positron.

(h) Map the field flux lines created through the spacetime curvature of an F-Sphere and note that flux lines create a magnetic field and dipole

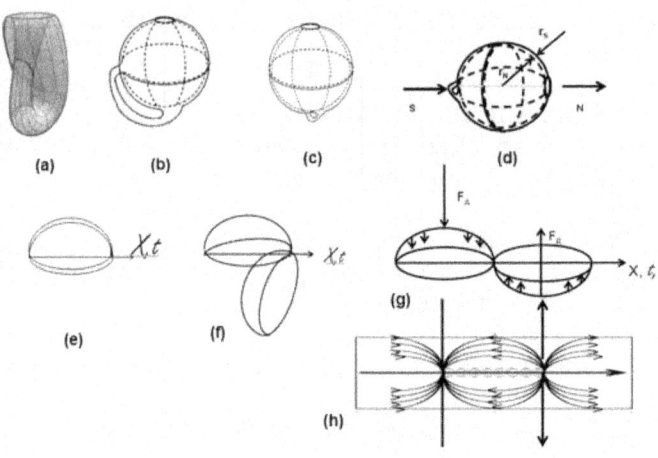

What's an electron in a metal matrix?

Moving on to a metal material, the not-so-free electron travel inside of a metal matrix is more like harmonic hopping, resonant vibrations from atom to atom one right after the next. Analogous to an acoustic shock wave through hanging colliding spheres of a Newton's Cradle. This may better be understood physically at the atomic level as a soliton hitting a network, vibration hopping from one atom to another, pushing a resident harmonic out to hop to the next atom in the network. Thus its frequency hopping travel is impeded with network or lattice atoms that are perturbed "mechanically" through essentially atom surface vibration of spacetime curves, otherwise measured as heat or temperature. The "resistance" to electron transport in a metal matrix or network increases with temperature because atom surface vibrations produce fluctuating spacetime acceleration-deceleration fields countering the passing spacetime curvature of electrons, which slows the electron's migration through a lattice. In superconductive material, the matrix creates a resonant network of atoms/ molecules that do not vibrate in a mode that impedes electron harmonic hopping.

A Newton's Cradle toy is a useful physical analog on electron transport in a metal matrix, where a swinging sphere on a string impinging on a row of sphere masses suspended in a row, metal lattice, will transfer force from an impinging swinging metal sphere to the last sphere mass in the row almost instantly propelling the last sphere from its pendulum rest position up to the potential of the initial or impinging sphere. In like fashion, the electron is a spacetime curvature harmonic bouncing to an adjacent atom will cause that stable atom to in turn to discard that extra harmonic to a neighboring atom and on it goes to the terminal end atom whereby the "electron" gives up its moving wave kinetic energy in the lattice to spacetime curvature configured for travel outside of the lattice.

What causes Superconductivity and how?

In the phenomenon known as superconductivity, scientists claim that electrons pair up in ways that allow them to travel without resistance. In conventional superconductors, electrons pair up only indirectly, as a by-product of the interplay between the particles and vibrations in their atomic lattice. Electrons "ignore" their fellows, but end up thrown together in a way that helps them to navigate without resistance at temperatures a few degrees above absolute zero. But in unconventional superconductors — many of which carry current with zero resistance at closer to 140 Kelvin — electrons seem to pair up through direct and much stronger interaction.

Superconductors have characteristic limiting temperatures below which the material's resistance drops suddenly and abruptly drops to zero and there is a complete ejection of magnetic field lines from the interior of the superconductor during its transitions into the superconducting state. This cannot be explained by simple perfect conductivity in classical physics.

It was discovered that some cuprate-perovskite ceramic exhibit these superconducting properties at temperatures up to 90 K. Such a high transition temperature is theoretically impossible for a conventional superconductor and increases their potential applications enormously. The classical solid-state lattice structure is shown below.

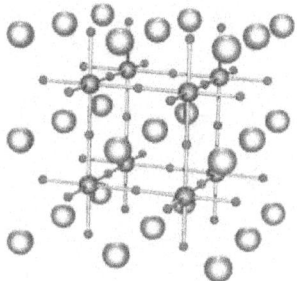

The structure of perovskite is shown with a general chemical formula ABX_3. The small spheres are X atoms (usually oxygens), the midsized spheres are B atoms (a smaller metal cation, such as Ti^{4+}), and the large spheres are the A atoms (a larger metal cation, such as Ca^{2+}). This structure somehow provides a spacetime curvature mystery that provides the above-mentioned superconductivity properties.

A layman's explanation using the Chronature model is, above "superconducting transition temperature" conditions, an atom impinging electron will perturb the atoms current surface boundary harmonics which will add the impinging wave and then seek to transfer it to a neighbor to retain a more vibration stable configuration. What is intuitively obvious and the classically accepted explanation was that when the atoms in a lattice remain in their respective lattice sites without random migration from site to site due to local inter-lattice spacetime changes, aka temperature due to kinetic motion from atom collisions in the lattice, overcoming lattice site bonds, electron transport is attained without extra atomic collisions. But how does this create superconductive properties without some kind of compelling force within the atom lattice or otherwise phase structure? Chronature models this as a creation of acceleration channels formed where atomic rotation, spin, or vibration at sites are allowed while site translation is slowed, all due to the spacetime configuration of the atom location, atom size, atom orientation, and harmonics. This restriction enhances translation through the lattice channels formed via the curved spacetime channels relatively clear of any spurious collisions in channels unimpeded to election flow, intra-atom harmonic shifts. Shown below is a Minkowski diagram of 6 atom lattice sites where the straight lines through the atom sites "straightened" to show their comparative length to the channel between atom rows. The channel spacetime continuum lines appear shorter at the channel centerlines. This creates the spacetime curvature, just like in a gravitational field, will rotate an entity coordinate frame such that the curvature component in the spacetime will appear, at the expense of a shortened local frame rotation forming an acceleration field. The acceleration field is maximum in the channel center, equidistant between the atom site rows, as it is additive $a_1 + a_2$ shown for two separate superposition acceleration fields, one from each of the adjacent rows, in the figure. This property forming acceleration channels in an atomic structure is proffered as the Chronature model for zero electron resistance to transport at superconductivity conditions. Please note that this is time v space but the atom lattice locations are determinative to time, ie at different spatial axis positions, the temporal lines do not line up in the same amplifying fashion.

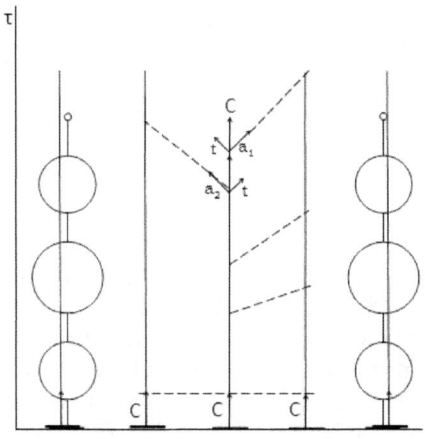

What is not explained by these hypothetical acceleration channels is the expulsion or ejection of magnetic field lines from the interior of the superconductor. Here the configuration, spin, rotation, the orientation of the individual atomic spacetime surface combinations at the sites determine how the magnetics will manifest, for conduction or repulsion in the superconducting matter. Using the Chronature F-Sphere model, each atom, with certain spin, rotation, and or orientation parameters, provides a magnetic dipole, called a Bohr magneton in its most elemental state. These are currently labeled "phase transitions." Hence at certain F-Sphere atom lattice configurations or material states, these atomic variables, in concert, give properties of superconductivity with the magnetic expulsion and pinning manifesting states, and all without the need for "Quantum effects." Furthermore, the related property superfluidity, where the viscosity of a material vanishes, is the macroscopy manifestation where the atom sites no longer rigidly cling to adjacent neighbors and configure themselves in such a way such that their local adjacent spacetime is no longer curved relative to each other, having no "bonds", whereby the channel spacetime is curved in parallel synchronicity.

It is well known and old physics that a magnetic field is created whenever an electric charge moves. Therefore, a free electron has an intrinsic magnetic moment associated with it as it moves. This is called the Bohr magneton (μB) and is a physical constant that is used in atomic physics to describe the magnetic moment of an electron in its ground state. Because lattice electrons are spacetime harmonic entities, higher harmonics represent higher electron's magnetic moments which are normalized to the magneton of an electron in its ground state. Hence, an electron's magnetic moment has to be an integer multiple of the Bohr magneton. The classical physics formula for the Bohr magneton in SI units is

$$\mu B = e\hbar/2m_e$$

where $-e$ is the charge of an electron, $\hbar$ is the reduced Planck constant, and I am the rest mass of an electron. Hence any understanding of superconductivity will need to address the physical spacetime effects of the magnitude and orientation of the atom magnetonage.

Known as the Meissner effect, under a materials transition temperature, a superconductor will cancel or expel any magnetic flux within its lattice. Moreover, many superconducting ceramic materials, aka the high-temperature superconductors, have perovskite-like structures. So this gives us more clues on how to find materials with relative lattice site locations, spatial, and harmonics or element electron configuration, temporal, for having material transition temperature required for superconductivity. By combinations and compounds of disparate materials, scientists have been able to obtain superconductivity at well above absolute zero, but still very cold, temperatures just above the important liquid nitrogen temperature. Type-II superconductors are usually made of metal alloys or complex oxide ceramics. Examples include the cuprate-perovskite ceramic materials which have achieved the highest superconducting critical temperatures. Due to strong magnetic vortex pinning, the cuprates are close to ideally hard superconductors. Currently, strong superconducting electromagnets often use niobium-titanium or, for higher fields, niobium-tin. We proffer that for these and other combinations of elements of differing size and harmonics at higher temperatures, synchronously maintain the acceleration channels open or at least re-route the channels around translating atom sites.

The Nobel Prize-winning physics theory on superconductivity, Landau–Ginzburg, is a mathematical physical model-based. Other theories progressed into more macroscopic scales given general geometric settings, placing it in the context of Riemannian geometry. So here we note that again the true nature of matter is modeled by the application of purely geometrical features through the use of non-Euclidean geometry. However, superconductivity theory is still not there. That is to say, it is not useful for predicting which materials and combinations will provide higher temperature superconductivity properties. They are explanatory but not predictive. Hence all current superconductivity material discoveries are made through trial and error, mixing, and testing materials for their superconductivity properties. What is needed are predictive theories, so that we can exploit one of nature's free lunches in applications saving energy and reducing damage.

Separate from the data and models on superconductivity, recently experimental data showed that twisted graphene at specific angles, 1.1°, unexpectedly exhibits the superconductivity property. Graphene is one of those especially useful materials and holds much promise for tremendous new applications because of its amazing properties. Since the graphene is a strictly 2D lattice, our speculation with regards to the Chronature model would posit that the atomic lattice is structured in such a way that the spacetime lines form an acceleration field permeating the lattice and channeling electron transport through the acceleration field as a body would "fall" in "free-fall" in a 3rd spatial direction. The magnetic dipole alignment is also set for maximum no internal self-cancellation of fields. Hence there are no "energy-wasting" collisions in the traveling electron curvature harmonic from atom to atom and no deceleration fields set up between lattice sites to the movement of the traveling spacetime wave in the lattice.

What's beta decay?

What about a physical description of beta decay, from which one of the two free, β⁻ and β⁺, electron or positron can be formed? Our model of the free electron is adjustable to explain the physical manifestation of the positive and negative "electron", wherein a violent breach of the atomic spacetime boundary produces these entities that we measure.

For this explanation, we go back to the morphed 4D Klein bottle wherein the one-sided surface or skin is oriented in one of two ways, with the outside as negative curvature or flipped to the positive curvature portion of the bottle. The twist in the bottle throat section between the two physical manifestations produces a dimensional rotation flipping the sign of the measurement by rotating the negative curvature and the positive curvature of the same-sided surface. You are probably thinking this physically cannot happen because bottles do not flip the inside of the bottle to the outside or vice versa or the curvature sign will change. You would be forgetting that it is the same side surface of space with a changing time, hence the lines of spacetime go opposite each other in time as we traverse the spacetime of the bottle. This is shown below, please say hello to an electron-positron pair. Note the different orientations produced by the throat reorientation between opposing curvatures. We could sure use MC Escher right about now, as you may be slowed with my poor artistic talent and need your imagination to carry this concept across.

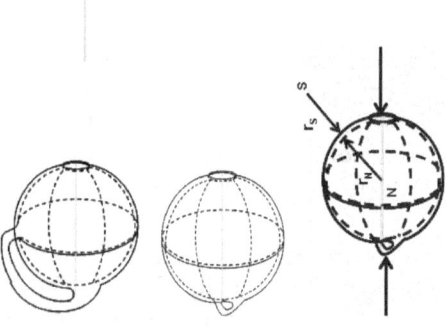

What we see is that the curvature from the inside can be flipped with the curvature of the "outside" and appear to anything else on the "outside" that the sign has changed, what we measure, hence the atomic "electron" has become a positron by a mere flip of its physical geometry. We proffer that this flip or twist in time is what causes the "backward in time" manifestation of the electron.

Why does the outer electron shell "want" to be full or empty?

An electron shell is the set of allowed harmonics for a particular atom, which share the same principle vibrating sphere harmonic number, n (the number before the letter in the harmonic label), that those spherical harmonic modeshapes can take on. An atom's nth "electron shell" can accommodate $2n^2$ spherical harmonics, e.g. the first shell can accommodate 2 harmonics, the second shell 8 harmonics, and the third shell 18 harmonics. The factor of two arises because the allowed harmonics are doubled due to the spherical symmetry of opposing physical axes; each atomic spherical harmonic admits up to two otherwise identical harmonics with opposite atom boundary modeshape lobes.

Currently Bohr model "subshells" are a grouping by the convention of "electrons", those whirling dervishes orbiting a nucleus, in progressing order of atomic number, number of protons. For us, a subshell is the set of atomic vibrating sphere harmonics defined by a common azimuthal harmonic number obtained from the vibrating sphere wave harmonic solutions, ℓ, within a group of similar spherical harmonics. The values ℓ = 0, 1, 2, 3 correspond to the *s*, *p*, *d*, and *f* labels, respectively. The maximum number of spherical harmonics defined by the spherical vibration in a subshell is given by $2(2\ell + 1)$. This gives two harmonics in an s subshell, radial harmonic, six harmonics in a p subshell, azimuthal harmonics, ten harmonics in a d subshell, and fourteen harmonics in an f subshell, more azimuthals. All the same subshell harmonic solutions from the vibrating sphere.

This vibrating sphere model solutions are analogous to the solutions reached by Schrodinger in the application of the Bohr model coupled with QM and Bohr's decree for a "quantized angular momentum of the electron" The harmonics will occupy the harmonic solution frequencies that most closely resonate with the atom boundary structure natural harmonics which are excited by the internal forcing functions, one degree of freedom per proton.

Quantum Numbers and Atom Curvature

Using QM solutions to the vibrating sphere, we obtain a set of mathematical equations, called wave functions (□), which QM-Bohr describes as the probability of finding electrons at certain energy levels within that region bounded by the wavefunction. In the Chronature model, the "electrons" are harmonic vibrations on a sphere whose 3D curvature in space and time is the quantifying attractive force accelerations, not the volume bounded by the wave function.

A wave function for an electron in an atom is called an **atomic orbital**; a 3D surface spacetime curvature within an atom boundary is the result of a spherical harmonic changing from a surface curvature with one frequency to a spherical surface curvature with a different frequency, usually accompanied by the absorption or emission of a photon, self-propelled spacetime curvature.

Each spherical harmonic in an atom is described by four different **quantum numbers**. The first three (n, l, m_l) specify the particular harmonic modeshapes of interest, and the fourth (m_s) specifies how many spherical harmonics can occupy that harmonic modeshape.

Principal Quantum Number (n): n = 1, 2, 3, ..., ∞

Specifies the **radius** of the first spherical modeshape, the distance from the nucleus to the sphere surface, in a group. All harmonics that have the same value of n are said to be in the same **shell**. For a hydrogen atom with $n=1$, the modeshape and harmonic is in their *ground state*; if the harmonic is in the $n=2$ modeshape, it is excited. The total number of harmonic modeshapes for a given n value is n^2.

Azimuthal angle modeshape Number: (l): $l = 0$, ..., n-1.

Specifies the **shape** of an orbital with a particular principal spherical harmonic number. The Azimuthal quantum number divides the shells into smaller groups of harmonics called **harmonic modeshapes (sublevels)**. A letter code is used to identify l to avoid confusion with n, where l is consecutively increasing lobe starting from 0 and progressing to n-1.:

l	0	1	2	3	4	5	...
Letter	s	p	d	f	g	h	...

The harmonic modeshape with $n=2$ and $l=1$ is the 2p modeshape; if $n=3$ and $l=0$, it is the 3s modeshape, and so on. The value of l also has a slight effect on the radial component of the harmonic modeshape; the harmonic frequency of the harmonic increases with l ($s < p < d < f$).

Orientation Number (m_l): $m_l = -l$, ..., 0, ..., +l.

Specifies the **orientation in space** of a harmonic modeshape of a given harmonic (n) and lobe shape (l). This number divides the harmonic modeshape set into individual **modeshapes**; there are 2l+1 individual harmonic modeshapes in each harmonic modeshape set. Thus the s harmonic modeshape has only one modeshape, the p harmonic modeshape set has three modeshapes, and so on.

Writing Spherical Harmonic Configurations

The progression of the harmonic frequencies among the modeshapes of an atom contributes to the atomic 3D surface **configuration** and, in theory, the magnitude of non-symmetric or non-balanced spacetime curvature. The harmonics comprising the atom's surface are filled in according to a scheme known as the **Aufbau principle**, which corresponds, for the most part, to the increasing frequency of the harmonic modeshapes:

1s, 2s, 2p, 3s, 3p, 4s, 3d, 4p, 5s, 4d, 5p, 6s, 4f, 5d, 6p, 7s, 5f

The order in which the harmonic modeshapes are filled into a vibrating sphere can be read from the periodic table in the following fashion:

Table of Spherical Harmonic by Quantum Number

n	l	m_l	Number of harmonics	Mode Set Name	Number of modes
1	0	0	1	$1s$	2
2	0	0	1	$2s$	2
	1	-1, 0, +1	3	$2p$	6
3	0	0	1	$3s$	2
	1	-1, 0, +1	3	$3p$	6
	2	-2, -1, 0, +1,	5	$3d$	10

		+2				
4	0	0		1	4s	2
	1	-1, 0, +1		3	4p	6
	2	-2, -1, 0, +1, +2		5	4d	10
	3	-3, -2, -1, 0, +1, +2, +3		7	4f	14

What does completing or emptying "outer shells" have to do with "bonding" and what is the nature of "valence electron" or atomic boundary harmonic resonance bonding?

"Bonding" is nothing more than the attraction between atomic surfaces taking into account the 1) attraction, acceleration field, from curved spacetime surface modeshapes at harmonics in resonance, 2) impulse repulsion from differing frequency, exclusion and times boundary interactions, time invariance 3) resonance frequencies of atoms in proximity, eg diatomic and covalent bond partners 4) average sphere surface curvatures, average spacetime curvature from the atom boundary, for non-resonant atomic harmonic frequencies, eg ionic bonds, and all the combinations.

When we speak of attractive-repulsive resonance of spacetime boundaries of atomic or molecular entities, we are referring to something akin to "locking frequency", but phenomena as applied by the forces and surface acceleration fields from the interaction of curved spacetime harmonics in resonance between atomic space curvatures at disparate atom local internal frequency-time harmonics. These may act as single or multiple degrees of freedom coupled oscillators in their interaction. This leads to a coupling of the oscillations of the individual degrees of freedom. This phenomenon was first observed by Christian Huygens, where two pendulum clocks of identical frequency mounted on a common wall will tend to synchronize or lock on to each other's harmonics. Other examples from macrophysical phenomena can be found in the disparate fields of nature and include such things as Sagnac Affect lock-in, gravitational locking, quantum levitation or quantum locking, Phase Lock Loop (FLL), phase lock frequency, mode-locking in lasers, Strouhal Affect in fluid-structure mechanics – producing vortex shedding in a fluid where there is a resonance between a fluid *vortex* shedding frequency and the structural natural frequency of a fluid impeding structure in the fluid path flow.

The atom boundary is very flexible; otherwise, we would not measure all of the properties as a result of the harmonic modeshape atom surface curved spacetime that we do measure. Being a flexible surface means that the surface has a thickness that is topologically stretchable, as is any spacetime, resulting in curvature. The surface boundary is where two spacetimes meet, having different time flows competing to maintain their geometric uniqueness.

What is the shape of an Electron?

According to Quantum Field Theory (QFT), the electron has no height, width, or physical measurements; it is a dimensionless particle orbiting the nucleus. However, by the best current model, the electron itself can be found in a cloud of virtual particle space. That "cloud" volume should have a measurable shape. The shape of that "cloud" is currently a highly sought, but unanswered question in Physics.

Along with the many questions in just basic fundamental physics, what is the shape of an electron, proton, and neutron? On these, there are generally many hand-waving flat-out WAGs based on unverified models or mentally exhausting breakdowns of the current state in Physics.

The current "best" model that physicists use is the Standard Model of Particle Physics. The mathematical framework for the Standard Model of Particle Physics is Quantum Field Theory (QFT), or I think of it as quantum fortuitous theory. According to QFT, an electron is a dimensionless point that has an associated field. That's the "wave" part. Anywhere in space, at any time, there are always particle/antiparticle pairs magically, remember in QFT the human mind is incapable of this understanding, popping in-and-out of existence; and, by QFT, the particle's field leads to a sea of virtual particles surrounding the dimensionless particle. Physicists talk about "measuring the electric dipole moment" of an electron to get a clue on its shape. The model to which they are referring seeks the asymmetry or the spherical nature of a swarming "cloud" of virtual particles. At some point, physicists don't separate the two, the dimensionless electron from the "cloud" of virtual particles. QFT'ers quickly start to treat virtual and physical as the same thing because that way they don't have to address the magical anymore. And QFT'ers offer lots of hand-waving arguments as to why this is so. Some QFTers contend that the quantum realm is so obtuse that our limited minds cannot possibly fathom the reality of the electron's true nature. Where have we heard that argument before, that all we need is just to have faith that in all its random nature the quantum model knows what it's doing?

Current theories leave us in a where's Waldo with the shape of an electron hinting it's in a "sea of virtual particles surrounding the dimensionless point of an electron," and therefore physically unfindable at least in a portrait. Moreover, the Standard Model offers that the shape of an electron should be generally spherical or at least, that any distortions in the electron's shape will be far, far too small to detect with current technology. But if we look at the photos taken of atoms and molecules developed at IBM and other labs, we see no electron orbital, we see not particles orbiting the nucleus, we see only some kind of boundary enveloping the nucleus.

In the language of the Standard Model, physicists hold that the electron has a near-zero electric dipole moment (EDM). The EDM is a measure of how "squished" a charged body is or a measure of how spherical an object is or is not. An EDM of zero defines the object of measure as a perfect sphere. Yet there is no dispute among physicists that this electron physically orbits the nucleolus in some fashion.

How does nature create a free electron, E field, B field, and dipole from spacetime alone?

The chart below starts from a one-sided Klien bottle, the vibrating spherical boundary of an atom boundary, and unwraps that one-sided bottle boundary such that the two equal and opposite spacetime curvatures are exposed to the regular 4D spacetime we live in, that is with monotonic time. What remains is GR, where the opposite spacetime curvatures create the attractive-repulsive fields that are the basic characteristic of magnets, EM fields, dipoles, E-Fields, B-fields that we currently model in a real physical world as scalar and vector quantities.

How nature creates a Magnetic Dipole ....

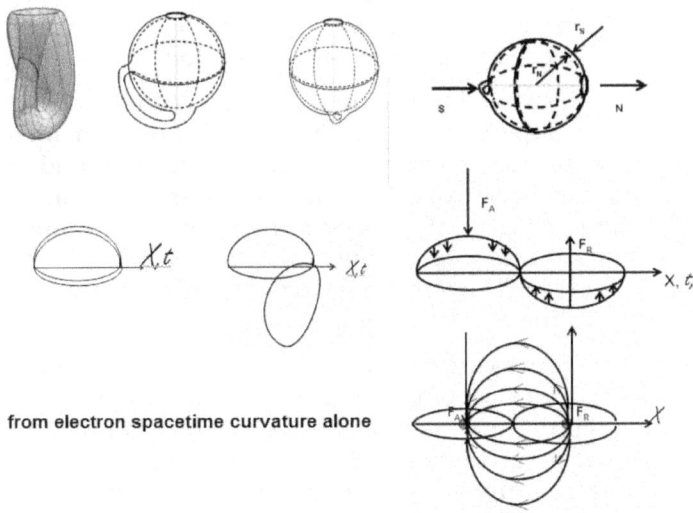

from electron spacetime curvature alone

Experiments with ytterbium monofluoride (YbF) molecules between electrified plates, and lasers to measure how molecules twisted, presumably gave physicists the shape of the electrons, within error, which suggested to them that the electron is spherical. However they admitted that their experiment/equipment just wasn't precise enough to know for certain, they could only offer an upper limit. A newer experiment had roughly the same setup, except the used ThO molecules instead of YbF. But, this gave scientists about an order of magnitude better result and they found that the EDM of the electron is still basically zero. All that work and no dipole.

What this means is the Standard Model already basically says any asymmetry in the shape of the electron is going to be too small, by many, many orders of magnitude, for them to distinguish with current equipment. In other words, the experiment still isn't sensitive enough, by far, to get down to the level needed to put the Standard Model to the test.

Related to this is the debate about a theory called Supersymmetry. The theory of Supersymmetry suggests to a physicist that the electron should vary eccentric or warped, much more than the Standard Model predicts. If Supersymmetry is correct, and the electron was "highly" warped, physicists would need to start looking for a vast array of other particles to add to the Standard Model.

Any variation in the electric dipole moment, no matter how small, will have huge ramifications for explaining the history of the Universe and the evolution of particles. Such as why are we in a matter-dominated, instead of light-dominated, Universe? What is the shape of the electron and what part does the EDM play? What ever happened to world domination?

The importance of the 3D physical and temporal shape is that it may have bearing on its binding affinity. Electron bonding theories abound. There seems no end to correlations in the electron bonding arena with correlations and correlations upon correlations. To mention some, physicists have given us the single-determinant approximation, Coulomb correlation, Born–Oppenheimer approximation, correlation energy, HF approximation, the correlation between electrons with parallel spin, the Fermi correlation, Coulomb repulsion, correlation related to the overall symmetry, dynamical and non-dynamical correlation, the configuration interaction method, static correlation and the degenerate determinant, Tomonaga Luttinger liquid model, bosonic interactions, correlated and uncorrelated pair density, to mention a few of these spinning nucleus electron orbiting particle theories. It's no wonder that physicists don't fully understand atoms, bonds, and molecules. The basic physical models aren't there to help scientists and so need cries out for creative speculation however physically implausible, in justification to proceed down the path of finding correlations, where a direct causal model cannot be imagined.

Linus Pauling hypothesized, that the lobes of the electron quantum wave mechanics conjured "probability clouds" were in some way responsible for the bonds between atoms and molecules. But Pauling was saddled with the Bohr electron orbital mechanics model with "help" from QM, and rather than shrug that off as Einstein did, Pauling tried to force-fit the Bohr atomic model with the chemistry of bonding. His concepts of hybridization laid the foundation for a century of speculative correlations and thinking for chemical bonds.

Perhaps not all of that speculation is useless. Pauling's idea that the electron orbital lobes were responsible for bonds was inspiration born without any good building blocks. In the Chronature model, the electron orbital lobes are dynamic curved spacetime and the lobe shape correlations can be used to verify chemical and electrical properties, a real and do-able result of spacetime curvature. The geometry of curvature calculations will verify physical measurements of the electronegativity scale made in the pre-spirit of hybridization bond characteristics. Furthermore, correlations between measured electronegativity and spacetime curvature attraction calculations can be used to validate computer codes. Calculations validating experiment results is the final reality check to assure that we do indeed have a valid atomic model and that we truly understand the real physical structure of matter.

To reiterate, currently, the chemical nature of the atomic bonds is not physically understood or explainable, hence the motivation for all the present creative theories on atomic bonding is mostly in using the electron shell "energy". In distinction, the notion that vibrating spherical harmonic mode shapes represent a family of spacetime shell curvatures, and spacetime curvature boundaries of discrete curvature, change between two spacetimes at characteristic harmonics of the atom spherical membrane boundary between monotonic time and circular time-space curvature, is pursued strictly on a physical model of an atom with positive curvature space and frequency-time, PCSFT, on the inside of an atom. The Chronature model holds that thin shell mode shapes are the "electrons" which are organized by increasing higher harmonics of spherical modes comprising spatial and temporal harmonics of a 3-sphere. The actual spacetime curvature, at any outside monotonic time boundary between the spacetimes, is a sum of the vibrating sphere mode shapes, analogous to the sinusoidal harmonic comprising any waveform in a Fourier series. These spherical harmonics of spacetime curvature are what are exchanged "electrons", which is also why electrons can only occupy certain "energy" levels or unique individual mode shapes of spherical wave harmonics, ie, we can only get integer number discrete harmonics, in or out. This is typically illustrated on our macro scale by strobing a vibrating rope to show only discrete harmonics, that "strobe" in the atom is the mode shape of the boundary between the two spacetimes, which when in resonance with a particular spacetime harmonic, will "pluck out" a harmonic or mode shape and transfer only that piece of the atomic spacetime curvature across the boundary containing only that resonance, the relaxing of one curvature spacetime to the other at time speed c.

To put this in perspective with a simple example using the current electron orbital jargon, a $1s^2\ 2s^2\ 2p_x^1\ 2p_y^1\ 2p_z^0$ p orbital Carbon atom can share a harmonic with a Hydrogen atom's 1s electron orbital by simply sharing the harmonic frequency of their atomic spacetime boundaries through resonance, whose boundaries would be mutually attracting interacting spacetime curvatures at those resonant frequencies as they share the same curvature across their interacting atom boundaries.

The atom boundaries at resonant frequencies bounce back and forth. The current "energies" calculated for the electron "orbits" or in our model harmonic mode shape of spacetime curvature in Hydrogen are shown below. The atoms with surface boundaries at resonant frequencies bounce back and forth attracted and repelled solely by the modeshape surfaces of the resonant frequencies of the highest resonant frequency, not the surface boundary of the Fourier sum. This is also known as the "co-valet" bond.

    W.K. Clifford[30] speculated that the phenomena of electricity might be explained by period variations in the curvature of space. Mathematically, this spatial dimension harmonic decomposition is much like a 3D Fourier transform. This appears much like a physical conformation to the Poincare Conjecture, that any 4D surface can be morphed onto a closed 3D sphere for the "compact", all loops on the object. In lower dimensions, Fourier proved that any curve is comprised of a series sum of sinusoidal harmonics. Likewise, any closed 3D surface with time is comprised of a series of higher 3D characteristic spherical harmonic lobes or mode shapes. But mathematics has gone much further in describing what happens when spaces are transformed from non-Euclidian inside 4 manifolds to outside in another 4-fold. Circles and straight lines are a good start for proving, the preservation of curvature of a circle inverted. These operations are in the mathematical inversion, Reflection, and movement categories of topological geometric entities.

---

[30] W.K. Clifford, *The Common Sense of the Exact Sciences*

Classic Theory of Electron Shells

The figure below shows the basic atom 3-sphere classical "S" shells, "probability zones" for certain "S" electrons. But in the Chronature model, they are indeed the radial spherical harmonics and modeshapes of spacetime harmonics at characteristic harmonic frequencies, not the actual volumes between spheres as treated by QM as "probability clouds".

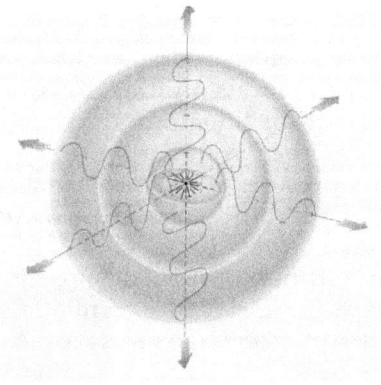

The formula from the classical Bohr model for "energy" between electron levels is $E = (1/n_1^2 - 1/n_2^2)$, where n is the integer number of the electron orbital level. With a decree that the electron particle orbit angular momentum was conserved, Monumentally fortuitous for Bohr his orbital angular momentum conservation model solution was a vibrating sphere equation. This stroke of luck gave QM proponents the "evidence" they needed to justify that the Bohr atom model was sound and useful. In contrast, using instead the difference in average spacetime scalar curvature between the harmonic modeshapes of a sphere to represent "energy", is approximated by

$$d(SpaceTime\_Curvature) = K(1/r_1^2 - 1/r_2^2)$$

where r_1 and r_2 are the average radius of curvature of the atom boundary modeshapes at harmonic 1 and 2 respectively before and after the atom boundary harmonic levels shift to transition some spacetime from within the atom to without. Instead of a Quantum number explanation, this yields the same formulation as derived by classical electron shell mechanics of orbiting particles. Thus the photon released as a result of changed "electron levels" or atom harmonic frequencies yields the transfer of spacetime curvature from one spacetime to another, which is conserved spacetime. Thus it is spacetime curvature that is neither created nor destroyed and conserved. Although we have not defined what "energy" or "mass" is in our model, we will relate, translate and verify Chronature elements to what is known about measurements of our physical world reality. But to reiterate, direct conversion between mass and energy is obtained through changes in spacetime curvature from one spacetime to another harmonic or another spacetime. Mass is the spacetime curvature configured in and by the atoms, and energy is the conversion of that curvature at discrete harmonics between two or more spacetimes or harmonics/modeshapes of spacetime.

Both Chronature and QM models provide the same results in this one aspect of "energy" transitions of EM and in predicting "electron shell" structure, AKA the transition of natural frequency harmonics of a vibrating sphere. This is currently known as Compton Effect.

Moreover, both Schrodinger and Chronature models rely on the inverse square law, $1/r^2$ outside the atom boundary for effects at a remote point location. This is more fundamentally understood as Gausses law of the E-fields. This coincidentally is the same mathematics for calculating the curvature of simple spherical 3D surfaces. Hence, this introduces more possibility of confusion, since the influenced point position is the same as the curvature of a sphere at that same point. The similar results in some cases between our physical spherical vibrating membrane model and the fudged QM electron orbital model now only muddy the waters as to QMs validity.

To restate, in the Chronature model of the atom, "electrons" are harmonics comprising the curvature of the atom boundary much like the Fourier sine and cosine curves comprising a curve locus, the atomic membrane is a thin divider between two spacetimes. The atom can comprise higher harmonics which contribute to a more convoluted spherical shell boundary, and these are defined as other "electrons" but are only higher spherical harmonics from more proton internal cyclical forcing functions resulting in an added curvature in a spacetime with high resonant frequency content.

GR predicts that the higher the space-space and spacetime curvature, the stronger the attraction between dimension embedded entities. Chronature expands on this concept to the very smallest of entities of what is recognized as "matter". The currently accepted "force" of gravity is thought too much weaker than any of the other forces and hence a different phenomenon altogether, so EM, weak force, and strong force are thus believed not to be the same force as gravity. However, one look at the dimensions of curvature involved, the scalar space curvature of an atom is so much larger than anything possible in regular or galactic space, except for the black hole singularity, where the gravitational force can become large due to the immense amount of matter causing the spacetime distortion.

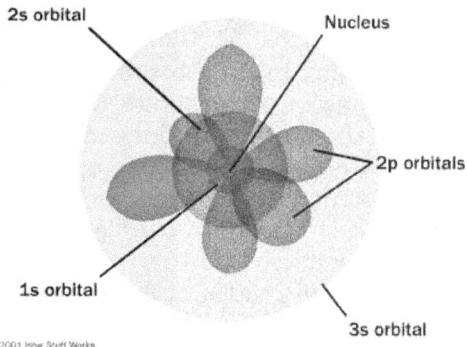

It should be noted here again, that it is not the classic orbiting electron "orbital" that is the electron variable, but the curvature of the spacetime which each harmonic adds to the atom boundary and the frequency of the harmonic which can with other species of spacetime curvature that effectively change the atom membrane curvature and through curvature attraction and repulsion "bond" and give "chemical" bond character through curvatures in resonance, spacetime.

Spherical harmonic mode shapes can produce very diverse spacetimes curvature, in space and in time. See the exemplar plot of the spherical harmonic surface below, for what would appear as "d" orbital harmonic modeshapes.

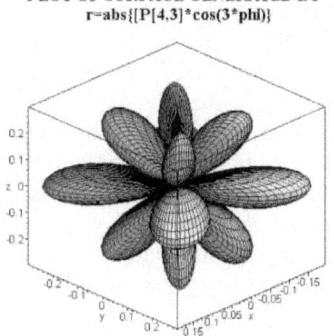

PLOT OF SURFACE GENERATED BY
r=abs{[P[4,3]*cos(3*phi)}

Why do the Schrodinger Equation and Quantum Mechanics non-physical atomic model accurately predict physical electron shell structure?

The explanation as to the physical interpretation regarding the electron shell structure provide by QM is what was impossible for Einstein to believe. The solutions from a probabilistic model [31] QM, predict the electron "orbital structure" for an atom but the physical interpretation is to a

---

[31] Donald H. Menze, *Mathematical Physics*, l. Pg. 212-213 "The significance of wave function. As far as the present volume is concerned, we shall not go into details concerning the relationship between ordinary vibrations and the atomic problems of wave mechanics. Nevertheless, the reader who may be familiar with the elements of the quantum theory cannot fail to see an analogy between the foregoing discussion and that of the classical methods of quantization. In the problem of the hydrogen atom, as originally discussed by Bohr and Summerfeld, arbitrary rules were employed to select, from a manifold of mechanically possible electron orbits, a few stationary states of special significance. The older atomic model postulated non-vibrating medium capable of sending out light waves. The model was definitely incomplete. The mechanical problem of selection certain stable modes from all possible modes of the vibrating sphere has many features in common with quantization. The parameters of m, l, and n are analogous to quantum numbers. The distinguishing difference is that in the mechanical problems the quantum numbers appeared naturally. We needed no special assumptions to introduce them, aside from the physical limitation that the amplitudes fulfill the boundary conditions and they be everywhere finite and continuous.

Schrodinger, in 1925, succeeded in applying the wave equation to atomic problems. Although he needed certain additional assumptions to adapt the fundamental equation to wave mechanics, the solutions follow in a manner similar to that already employed. The letters m, l and n, as introduced in the problem of the vibrating sphere, have been adopted because of their relation to the atomic quantum umbers. Indeed, the functions ψ and θ for the hydrogen atom are **identical** with those for the vibrating sphere."

probabilistic and **non-physical model** of the atom which was fixed by decree that orbiting electrons had a constant quantized angular momentum. Hence the physical inconsistency with orbital mechanics, constant angular momentum with wildly varying orbits.

The Bohr explanation for the wave equation solution's manipulation did not address the physical model of the atom and it was left up to QM to explain what couldn't be explained by any reasonable physical model. The wave solution to come out of the Schrödinger Equation came from manipulations of a "quantized" constant angular momentum Bohr model, and not to be confused by the non-wave solution producing Schrödinger Equation.

$$-\frac{\hbar^2}{2m}\frac{\partial^2 \psi}{\partial x^2} + V(x)\Psi = i\hbar\frac{\partial \psi}{\partial t}$$

Mathematically spherical harmonics are the angular portion of a set of solutions to Laplace's equation, represented by Y_ℓ^m with possible a forcing function model providing a radial component to a family of elements found in the Periodic Table.

As we shall see this is important in the Chronature model because the spacetime curvature of the atomic boundary, geometry of the spacetime boundary surface by itself, sets the physical, chemical, and electrical properties of an atom through a very flexible surface curvature of spacetime producing attraction or repulsion forces.

The gravitational, chemical, electrical, etc. forces as defined by GR as gravity, are a result of the number of protons bouncing around within perturb the atom boundary and excite the natural spherical harmonics from the lowest harmonic to the highest. This corresponds to the number of protons in the atom driving the harmonic excitation, the forcing function, of the boundary giving the element atom the known and measured properties that it has.

In contrast with QFT, where the same spherical harmonic solutions are derived serendipitously equal to the square of the Bohr model quantized electron orbital angular momentum operator by divine decree:

$$-i\hbar \mathbf{r} \times \nabla,$$

Quantized angular momentum was decreed by Bohr to represent the different quantized orbiting electron angular momentum by atomic orbital electron configuration. Moreover, the accuracy of the quantized atomic orbital electron configuration is what gave QM'ers their biggest proof that the Bohr model of the atom's electron shell structure was sound, or at least on the right path. We proffer that QM got the right answer using an incorrect model force-fit by decree from its author and major proponent Bohr. Bohr hypothesized that the electron orbits are quantized angular momentum states that somehow preserve the electrons in their physical orbits, and that "proved" the Bohr model of the atom was somehow correct albeit incomplete. It followed from this model that "electrons" were indeed particle satellites orbiting a nucleus in some unexplainable fashion.

To re-iterate because of its historic importance in changing the course of Physics, QM is correctly predictive of electron shell structure based on a badly conjured Bohr model with divined assumptions and QM density probabilistic wave equation, not based on physical reality or a physical model. QM proponents believe that the Bohr model is correct without additional explanation as to its relationship to physical reality. By QM we are to accept that physical reality rests on the probability of certain states manifesting like magic because an observer makes it happen. And therein lays the problem. To re-iterate, the spherical harmonic model is based on the physical atom being an undulating or vibrating spherical shell boundary separating two different spacetimes. This boundary of curved spacetime is comprised of spherical harmonics and the physical spherical shell boundary edge of two separate and different spacetimes. This boundary is dynamic and the harmonics of the spherical shell are determined by the number of forcing functions from inside the atom boundary, perturbing nucleons and protons causing the spherical wave forcing functions harmonics corresponding to the number of protons and equal to the "electrons" or harmonics.

    The degree to which the number of "protons", yet another variety of spacetime, cause the atom boundary vibration for the spherical wave to be symmetric or asymmetric, result in the spherical shell harmonics for neutral or "valence available" atoms respectively. Nobel elements in the Periodic Chart are the most symmetric from a spacetime curvature standpoint and hence not attractive or reactive, neutral. Where there is not asymmetric boundary co-incidence of physical attraction, we have atom surfaces that do not attract with and form "bonds" with other atoms as their harmonic modeshape symmetry produce a neutral surface over time.

Orbital Table

The electron shell structure then identically follows and uses the same "quantum numbers" as the harmonic solutions for the spherical wave or vibrating sphere, some solutions are shown below in a diagram from Wikipedia under the topic of Atomic Orbital[32] which shows a partial quantum number, (n, m, l), classification of the electron shell structure beginning with the fundamental mode or harmonics and progressing through the s, p, d, and f shells.

This table shows all orbital configurations for a hydrogen-like non-physical quantum wave functions up to 7s, and therefore only maps the simple electronic configuration to all elements in the periodic table up to radium. ψ graphs are shown with - and + wave function phases shown in two reflecting lobes. The p_z orbital is the same as the p_0 orbital, but the p_x and p_y are formed by taking linear combinations of the p_{+1} and p_{-1} orbitals (listed under the m=±1 label). Also, the p_{+1} and p_{-1} are not the same shapes as the p_0, since they are pure spherical harmonics. Bear in mind that this is only one harmonic classification system and is purely artificial. Just as important would be the frequency of each lobe and its classification.

---

[32] http://en.wikipedia.org/wiki/Atomic_orbital

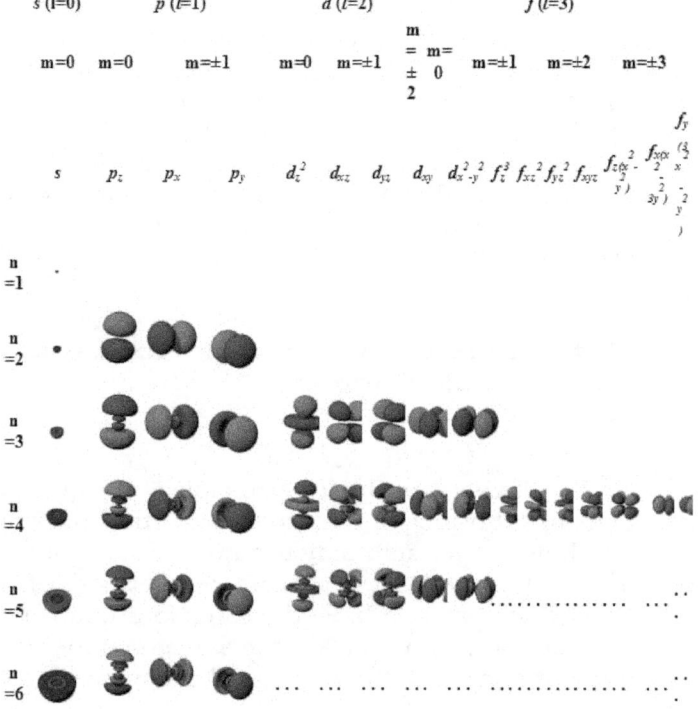

Figure ___

Shown in the above figure are visual representations of the first few **real** spherical harmonics. The **imaginary** spherical harmonics form yet another set not shown here. Red portions represent regions where the function is positive, and blue lobes represent where it is negative. The distance of the surface from the origin indicates the value of $Y_\ell^m(\theta, \phi)$ in angular direction (θ, ϕ). Bear in mind again that spherical harmonics are a series of special functions defined on the **surface** of a sphere, hence the solutions to the equation are a curved surface. Wherein QM claims that the lobe volume is where an electron probably exists at any time. So here again we see how the QM derivation of the electron orbital model is a forced fit from the supporting mathematics of spherical harmonics to achieve the same answer as measured for the electron shell structure.

What is the relationship of Pauli's Exclusion Principle to Chronature?

The Pauli exclusion principle is the quantum mechanical principle that no two identical particles with half-integer spin may occupy the same quantum state simultaneously. For example, no two electrons in a single atom can have the same four quantum numbers; if n, l, and m_l are the same, m_s must be different such that the electrons have opposite spins.

Since QM does not reflect a physical model of an atom or sub-atomic particles, but a probabilistic model, the explanation for what this means physically in Chronature is in order. With any electron subshell, orbital, or energy level, only two electrons can exist, one spinning clockwise and the other counter-clockwise is the current Pauli explanation.

By the Chronature model, we know that electrons are not orbiting particles nor do they orbit in any particular clockwise or counterclockwise direction. So what is it we are to make of basic repeatable observations, what interpretation are we to make about close electron subshell electron pairs?

Chronature is a physical model representing a physical multi-spacetime interaction model. In the case of an atom, all solutions to the Vibrating Sphere equation represent unique harmonics. By inspection, the mode shapes of certain harmonics travel in pairs, generally on equal and opposite sides of a common axis and opposite the atom's center point. One additional obvious observation is that all harmonics are unique, and go progressively from low to high frequency. The pairs of harmonics called a "shell" have the same "energy" because of the symmetry in the solution and are hidden in the numbers but physically manifest by inspection of the modeshapes. Hence in our purely geometric model of the atom, electron orbitals are harmonics that have 180-degree phase-shifted pair harmonics or mode shapes from the common axis about the center symmetry have the same "energy," ie have identical curvatures and mode shape opposites wrt an atomic center and common axis. See the figure just above for the family progression of pairs of electrons representing these Pauli subshells. See the figure below for the 3 p-subshells.

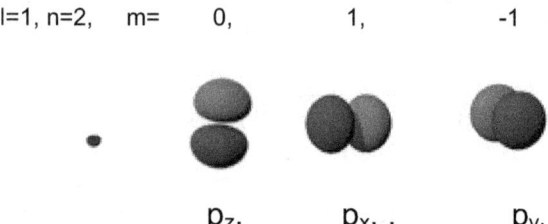

l=1, n=2, m= 0, 1, -1

p_z, p_x, , p_y,

     The data that Pauli relied on to reach his Principle provides Chronature with the current indication that "energy" at the "electron shell level" applies to the curvature of the same pair harmonics modeshapes of spacetime curvature of spherical vibrating membranes of spherical spacetime we call atoms. The modeshape pairs are shown above for the p subshells of x, y, and z. Additionally, this further embodies the correspondence for the term electron, between the solutions for a non-physical Quantum Wave Equation and the physical model of a vibrating spherical curvature shell proffered by the Chrono Spatial Dynamics model of the atom.

Photos at an Electron Exhibition

     In 2011, using an Atomic Force Microscope (AFM) IBM produced the first images of an electron's path. The images of the electron's purported orbital path around a nucleus are shown in the sequence of pictures below.

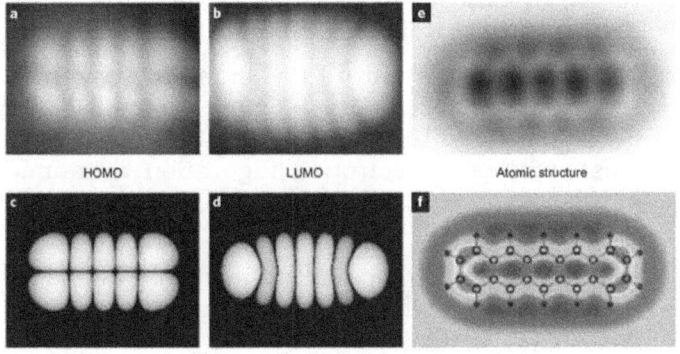

There was much discussion as to why the pictures did not manifest any resemblance to the current enhanced Bohr model orbiting electrons. Many offered entertaining solutions, nobody knew the answer as to why the Bohr model orbitals are conspicuously absent. The pictures of the electron; appear to bear out our physical model of the structure of matter from a visual point of view. That view asserting that inside these gray pillows something more uniform than an orbiting electron, something fundamentally different on a dimension scale. The follow on question is what is the substance of and the fabric of this pillow-like membrane covering of an atom?

How do pure geometrical differences in vibrating sphere modeshape curvature develop into physical properties of "Ionization Energy" and "Electron Affinity"?

The diagram below shows the current Period Element Chart trends in these measured properties.

Electron affinity – has much to do with symmetry, and elements get more symmetric as their harmonics tend toward the Nobel element row, the most symmetric electron configuration, in atomic number.

"Ionization energy" – increases with the element p, d, and f spatial harmonics. This happens because these higher modeshapes add sharper curvatures at the lobe ends and over smaller angles, more narrow petal end modeshapes, with increasing atomic number, and decrease with the element s shells because the larger radii reduce total curvature by $1/r^2$

Electron Knots, how is this physically possible?

Recently scientists at Aalto University and Amherst College created "knotted" solitary waves, or knot solitons by exposing a Rubidium condensate to rapid changes of a specifically tailored magnetic field. Physicists have been theoretically predicting that it should be possible to have "knots" in quantum fields and so a gas of superfluid atoms, also known as a Bose-Einstein condensate (BEC), is now offered as another proof of QMT usefulness in prediction. From Quantum Dots to Quantum Knots, QMers dutifully scurry down to attempt to explain physical observations from a purely probability-based non-physical model, no matter how sorely lacking in any reasonable sufficiency of a physical model which will comport with any believable real-world model.

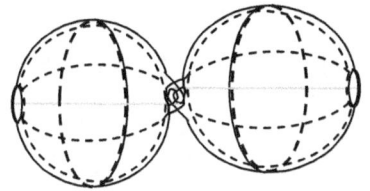 Current visualization of the structure of the created "quantum knot" shown above courtesy of David Hall. Each band represents a set of nearby directions of the quantum field that is "knotted." In theory, each band is twisted and linked with the others only once. But what is "knotted" and what does each "band" represent?

If indeed the physical atomic structure is thusly formed of this BEC structure fashion, the Chronature F-Sphere provides a compelling visual explanation of this BEC quantum picture. The visual can be made with a simple reconfiguration of spacetime entities with morphed Kleinian umbilicals in low external induced vibration. See below a physical manifestation for our happy couple in a knot.

5 – Nucleons and Nucleus Organization

Nucleons can occupy much larger dimensions inside the atom. This is because non-Euclidean geometrical dimensions extend from large to small as one progresses from the center to the atom boundary. As nucleons spend most of their time near the center due to the strong repelling gravity from the atom boundary to towards the center, strong force, the nucleon size may be much larger inside the atom than their measured dimension outside the atom in accordance with elliptical geometry dimension progression from zero, the center. However, the precise dimensions of nucleons are only approximately known for nucleons. Moreover, since we teach that what is known as "electrons" bounded to an atom are indeed truly 3-sphere harmonics on the atomic boundary, then what does make nuclear particles? The short answer is that nucleons are different spacetime 3-spheres inside the atom in which they reside. The nucleus boundary harmonic characteristic spacetime curvature is somewhat determined by nucleon properties, protons, and neutrons. Since the radius of a typical nucleon is ordered smaller than the atom, the nuclear spacetime curvature is corresponding orders larger than the atom's boundary hence nuclear "forces" and nucleon "masses" are orders of magnitude larger than atomic, chemical, electrical, and Van der Waals forces. This larger curvature in the nucleons means more powerful gravity forces which translate to higher "mass" as we measure the properties between the atom and the nucleons.

The main difference is between what is known as the proton and the neutron. The proton is not completely symmetric, which provides a spacetime curvature the is interactive with the atom boundary. Picture a grape seed bouncing around in a grape, pushing the grape skin around from the inside. The grape skin curvature pushes against the asymmetric seed, and the seed moves and pushes the grape skin back, acting like a simple harmonic oscillator, the proton excites only one atom wall harmonic per proton. Two protons mean 2 simple undamped harmonic oscillators and so forth. Hence a Hydrogen atom has one proton and thus one corresponding harmonic, "electron". Since the atom is essentially a 3- sphere, it has other possible harmonics that are not excited unless it acquires another proton, which we then classically label another element because the next harmonic, electron, changes the atom boundary spacetime curvature causing different electrical, chemical, etc properties.

The standard model identifies known discovered particle properties and labels these under the quarks, gluons, bosons, etc to the entities observed in atom smashers and measured by various means. But we proffer that these sub-atomic entities are nucleon boundary harmonics, properties, and transitions between spacetimes, analogous to EM between macro spacetime and atomic spacetimes. Thus quarks can transition to other quark levels, much like electrons, releasing or acquiring some spacetime very "high energy" curvature in faithful compliance with the conservation of spacetime curvature law stated above and spherical harmonics for which they are physically in similar analogy to electrons. Even their dimensions are similar. See the chart below, showing that the dimension of the quark and the dimension of the electrons are strikingly similar at about 10^-16 cm. This may be because both the quark and the electron are both spacetime boundary oscillations and the boundaries have the same thickness dimensions.

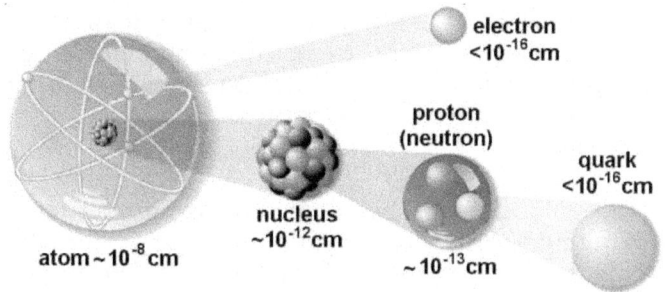

Of course, the Strong Force is credited with keeping the nucleons in their place inside the atom walls. But what is the Strong Force? In Chronature, it turns out to be just anti-gravity, a boundary curvature inducing repelling force, instead of an curvature attractive force, and towards the atom center. That's correct, a repulsive acceleration from a rotating universe, Gödel, time and non-Euclidean Elliptical geometry 3 space. Conceptually, the atom boundary is inverted facing the inside and opposite to the outside spacetime curvature, having an opposing elliptic geometry metric, so instead of attractive, it creates a repulsive acceleration field, compelling the entities inside towards the center. And because the geometry is non-Euclidean Elliptical, the spacetime grid is much steeper than gravity outside, thereby exerting a much stronger compelling acceleration field on nucleons corralling them to move around but only in the vicinity of the atom center.

The protons in the nucleus experience a repulsive electrical force due to another proton (s) having the same "positive" charge, but there's also the residual strong force to consider acting against the electrical force of charge. The Strong Force, SF, is much stronger than the electromagnetic force and keeps the nucleons within proximity of each other or "glued" in a region of space on a scale of femtometers.

Plus the SF has some strange properties like being non-central and including 3-body interactions. This is because the repulsive nature is what dominates generally inside the atom, and emanates from the atom boundary inside the negative curvature surface.

One of the things that makes the strong force "strong" is the fact that curvature is concentrated and focused on the center, which is opposite and diffusive of positive curvature space and time with diminishing spacetime radii of curvature away from the curvature line or boundary outside the atom. Rather than getting weaker with distance, like electromagnetism, which has infinite range but generally decreasing flux at the inverse of the square of the distance, the SF gets stronger with distance from the boundary as the negative curvature grows in the negative direction of curvature with smaller radii of curvature with distance from the curvature boundary, essentially "concentrating" the SF toward a center with distance from a spacetime boundary, instead of diminishing, as such creating a "potential well". This is much like a nuclear bomb design having all of the explosive force detonating at once so the compression wave set off simultaneously on the outside boundary will concentrate its compressive force exponentially as the compression wave flows to the sphere center.

For another example, the "color force" between quarks, explained as "potential gradient is constant and does not get stronger with distance", has spacetime curvature meaning that the spacetime curvature between quarks saturates or maintains a constant radius of curvature at some large constant value at large distances. Whereas the SF which acts between hadrons has a finite range, which is roughly inversely proportional to the pion mass, saturating at lengths of a few femtometers (10^{-15} meters) with a maintaining curvature with a different radius. The point here is physicists currently talk in terms of "mass" and "energy", which are physically measurable due to the spacetime curvature inside the atom and in the disparate spacetime nucleons.

Nuclear Spacetime Model vs. the Nuclear Shell Model

The Chronature nucleon spacetime model is analogous to the atomic spacetime model, with the distinction of a five order magnitude smaller spatial neighborhood and a negative frequency for the time dimension.

This is somewhat analogous with the current Nuclear Shell Model and the Schrodinger equation electron shell model substituting the arrangement of electrons in an atom with subatomic "particles" comprising the nucleons.

As currently accepted, electron shells that complete certain shell levels result in greater atom stability and non-reaction or neutrality characteristics. Hence these elements are given special status but the explanation is that they complete some kind of valance orbital. Chronature models are based on harmonic symmetries at these frequencies and we proffer that it is the symmetry of the 3-sphere produced, causing the curvature to be uniformly attractive from all directions with rotation, therefore "neutral", equally attractive in all directions.

Just as adding electrons to the atom changes electrical and chemical properties by virtue of changes in the atom's surrounding spacetime curvature, adding nucleons, protons or neutrons, in a nucleus provides certain combinations where the "binding energy", the measured mass contributed per nucleon, of the next nucleon is significantly less than the last one. But there are certain "magic numbers" of nucleons: **2, 8, 20, 28, 50, 82, 126** which are more "tightly bound" than the next higher number and provide a more stable nucleus. Strangely enough, the classic atomic electron shell model has similar "magic numbers" but where the "electron shells" or nucleon spacetime harmonics are at **2, 10, 18, 36, 54, 86,** ... and hence provide a more stable atomic element, the Nobel elements.

Z	Element	No. of electrons/shell
2	helium	2
10	neon	2, 8
18	argon	2, 8, 8

| 36 | krypton | 2, 8, 18, 8 |
| 54 | xenon | 2, 8, 18, 18, 8 |

| 86 | radon | 2, 8, 18, 32, 18, 8 |

In the conventional nuclear Shell Model, due to some "variations in the orbital filling," the upper magic numbers are 126 and, speculatively, 184 for neutrons but only 114 for protons. This has opened up for a theory playing a role in the search for the so-called island of stability, where the nucleon configuration leads to a stable nucleus.

Theories and explanations abound[33] on all the slight differences as to some of the magic numbers, experimentally and empirically established. These theories start from the shape of the nuclear "potential well" enclosing the nucleons from electrically repelling themselves apart, to "spin-orbit coupling" and others. These models are all based on empirical data, and subject to discontinuities and anomalies to the accompanying associated ancillary theories.

The current nuclear models and theories are based on a variety of parameters such as atomic number, quantum effects, magnetic dipoles, nucleon spin or parity, etc. But none of these have any real physical basis, as the Bohr model does not accommodate the physical, and QM is unreal, or probabilistic at best. But measured numbers and

---

[33] Hans A Bethe and Philip Morrison, *Elementary Nuclear Theory*, pg 162-173

metrics are necessary and helpful in designing nuclear reactors and nuclear devices but do not get us a physical explanation of adequate understanding.

To reiterate, as in the atomic shell model for electrons, nucleon spin or orbit models do not have a physical model, only a non-physical quantum statistical model understanding, just like in the atomic electron shell model. However, the Nuclear Shell model does provide analogous four quantum numbers to the atomic electron shell n, l, and m_l, m_s, called n, l, and m_l, m_s, nucleon shell number, nucleon angular momentum, nucleon spin, and nucleon magnetic quantum number. This should give us a clue, that like the predecessor Bohr atom model, the nuclear model of a nucleon is a 3-sphere having a different spacetime. All else being equal, the spin, parity, and other characteristics operate similarly for the nucleon spacetime, making changes to accommodate the differences in the negative frequency-time dimensionality. This is manifested as the very short life span of any nuclear matter outside the nucleus.

Although substantially different in kind, the commonality and analogous nature between the modern particle models and the Chronature models of spacetimes are too much to be merely coincidental. Moreover, the basic underlying similarity in structure of the atom and its nucleus, apart from the spatial dimensional scale, are also too accurate to be coincidental. That is to imply that the basic model for one must be the same as the other with a few minor differences. The Chrono spatial dynamics of spacetime curvature would appear to be scalable, allowing for the slight changes in the basic dimension of time and geometry to the 3-sphere model of spacetimes, to dictate the differences in properties at the boundaries between spacetimes. This would indicate that the boundary character of the nucleon, stress/density would be radically different to accommodate the much higher spacetime curvatures and reverse frequency-time.

The traditional building of a nucleus is by adding proton and neutron particles that always fill the lowest available levels first, just as in the atomic electron model structure. Thus the first two protons fill the first harmonic; the next six protons fill the second harmonic set, and so on. As with electrons in the atomic model, protons in the outermost shell will be relative "loosely bound" to the nucleus if there are only a few protons in that shell because they are farthest from the center of the nucleus. By this classical nuclear model, nuclei that have a full outer proton shell will have a higher "binding energy" than other nuclei with a similar total number of protons. All this is true for neutrons as well.

This would indicate that the magic numbers are expected to be those in which all occupied shells are full. This however does not pan out over a full set of magic numbers which means that this underlying theory of the "magic numbers" is not based on a complete vibrating shell harmonic model with positive time or positive frequency-time.

In the current Nuclear Shell Model, all nucleons, that is neutrons and protons, composing any atomic nucleus, have the intrinsic quantum property of spin. The overall spin of the nucleus is determined by the spin quantum number S. If the number of both the protons and neutrons in a given nuclide is even then $S = 0$, i.e. there is no overall spin. Then, just as electrons pair up in atomic orbitals, so do even numbers of protons or even numbers of neutrons pair up giving zero overall spin. These are similar to the atomic model electrons.

The 3-sphere Chronature spacetime model of the nucleons holds that the nucleus is also a 3-sphere vibrating boundary between 2 spacetimes, the atomic and the nuclear or sub-atomic. They differ by a negative frequency-time dimension in the nucleon spacetime. The spherical harmonics of this spacetime boundary are the quarks, with neutrons and smaller subatomic particles constituting the even spacetimes inside providing vibrations and perturbations to the nuclear shell boundary, much like the nucleus does to the atomic shell boundary. The bosons and gluons are much analogous to EM outside the atom, as when they are released from the nucleus they must transition to the entered spacetime.

We propose that the neutrons are 3-spheres of spacetime curvature with symmetric harmonics and the protons have asymmetric harmonics on this boundary, providing neutral or driving perturbations in atomic internal properties and atomic spacetime boundaries for one but not the other. The proton surface curvature contribution produces an asymmetry to the nuclear 3-sphere whose modeshape contains lobes with are attractive to the electron modeshapes of the atomic 3-sphere boundary. Thus each neutron or proton is comprised of sub-harmonics, which we currently have identified as quarks with various properties.

The Standard Model describing all the currently known "particles" contains six flavors of quarks (q), named up (u), down (d), strange (s), charm (c), bottom (b), and top (t). Quarks are spin one-half, particles, implying that they are fermions class. Since quarks are subject to the Pauli Exclusion Principle, which states that no two identical fermions can simultaneously occupy the same quantum state, then they are boundary harmonics, analogous to the atomic electron. From the atomic Chronature model we proffer that this indicates only that harmonics and modeshapes come in pairs, and a set of pairs can only accommodate two in the set. Putting these together points to a proton-neutron model is yet another type of spacetime with "s" and "p" type harmonics from the atomic model. The 6 quarks, up/down strange/charmed, top/bottom, fill the "p" shell with petal modeshapes in the x, y, and z-axis, one for each positive and negative axis. Other "subatomic" particles such as hadrons, baryons, gluons, etc exist as higher spacetime boundary harmonics in the same proton and neutron 3-spheres.

If the nuclear boundary is breached, the protons and neutrons behave as 3D solitons comprised of nuclear spacetime. Their half-life is generally short because their attraction is to resonances of similar spacetime curvatures found in other nuclei, similar from whence they came, and not the atomic boundary curvature which they pass.

Analogous to the current Standard Model particles, the Chronature models the nucleon harmonics from the solution for the wave equation of a sphere, where the spherical harmonics are created in the order of increasing vibrating sphere wave solution harmonic stable mode numbers.

In the atomic model, the noble elements have complete valence electron shells, the outermost electrons are normally the only electrons that participate in chemical bonding because their curvature contribution to the 3-sphere boundary is determinative over the last symmetrical contribution. Atoms with full valence electron shells, symmetric 3-spheres, are extremely stable because they are gravitationally neutral, and therefore do not tend to attract or be attracted, ie to form chemical bonds. However, "heavier" noble gases such as radon are held less firmly together by the spacetime boundary than smaller radius higher curvature noble gases such as helium, making it easier to remove the highest harmonics from heavy noble gases.

From a Chronature model, the Strong Force and the Weak Force must explain the manifestation of physical reality inside the atomic boundary spacetime curvature through spacetime curvature and the atomic modeshape harmonics.

A useful idea about the atom nucleus is "nuclear energy", and not just any energy, but the real physical and measurable "binding energy" between nucleons in some form of spacetime warpage. Why does this exist, what does it mean from the physical model standpoint, and how can we use it to further obtain a safe clean form of energy? The currently measured binding energy curve is shown directly below.

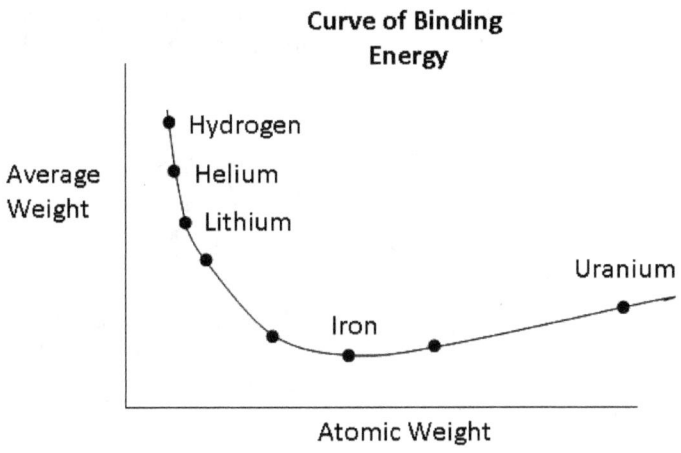

The two useful nuclear processes that we have so far been able to exploit, make physical measurements, and extract "energy" are the fission and fusion processes. Explosive exothermic energy arises from the strange fact that not all nucleons have the same mass that they had before the fission or fusion. As the Curve of Binding Energy shows, some nucleons are "heavier" than others. We know from $E = \Delta mc^2$, that the change in atom's energy, the energy produced from a fusion or fission, must come from the mass defect, Δm, the difference in mass from the reactants, and products. We opine that the mass defect is the difference in spacetime curvature from a first-order calculation, whereby the transfer of spacetime from or to the products renders their final states and products, as having less "mass" and hence the "mass defect". The "missing mass" is found in the spacetime geometric expansion and hence added spacetime curvature to the immediate reaction positions with the emission of products and EM.

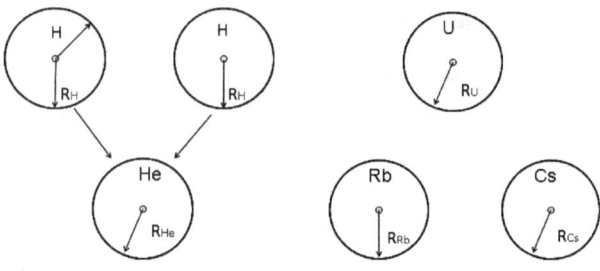

What is the mass defect and what causes the manifestation of a nuclear explosion?

Where does the mass defect come from, how it manifests, and what it is physically is a modern mystery lost somewhere in spacetime subatomic particle "bonds" and "nucleon binding energies". We all know that there is great kinetic energy imparted to the entire fragment particle set involved in an atomic rearrangement, seemingly all at once at fusion or fission. That is fusion or fission, coalescing or splitting spacetime boundaries. What is physically responsible for all of this resulting extremely large kinetic particle and matter rapid burn gas-like explosion? We know it is not a chemical expansion like a conventional explosive because the phenomena occur and are currently explained from inside the atom, not molecular action breaking of outside the nucleon "binding energy."

A nuclear "explosion" is a completely different kind of expansion than the expansion of explosions caused by chemical reactions. We proffer that the nuclear "explosion" expansion arises as a result of spacetime metric changes. Since the metric expansion is a key feature of the Big Bang cosmology, its currently accepted modeling and measurement are valid only on large scales roughly the size of galaxy clusters and larger. This is the current thinking mainly because gravitational attraction binds matter together strongly enough such that metric expansion cannot be observed at this time, on a smaller scale, as we would not notice the change on a small scale or in everyday life. This is made even more difficult during nuclear fission or fusion but mathematically there is no such limitation on spacetime expansion or contraction. But if look at a nuclear-induced expansion from a standpoint of pure kinetic energy and time, there are no other explanations other than spacetime expansion. On a large intergalactic scale, tests of distance and experiments show that space is expanding, even if a ruler on earth could not measure it, so we know that spacetime can expand is a consequence of atomic and subatomic reactions, at least in the far reaches of the Universe.

Going back to geometry of spacetime for a purely mathematical explanation of such explosively high kinetics, *Non-Euclidean Geometry*, Kulczycki[34] notes that under certain conditions, a non-Euclidean circle is nearly fifteen times greater than the $2\pi R$ of the Euclidean circle. He makes a similar argument

---

[34] Stefan Kulczycki, ***Non-Euclidean Geometry***, pg 170-172

and calculation for the area of a sphere between geometries. Kulczycki writes:

> "Observations similar to those made when discussing the circumference of a circle apply to formula 21.
>
> (21) $P = 4\pi k^2 * \sinh^2 R/k$
>
> The area of a sphere of radius R is greater in Lobatchevscian geometry than in Euclidean, and for R>k it is even much greater. If R = 5k the area of the sphere is about 220 times as great as $4\pi R^2$, which is its Euclidean value."

   Taking these mathematics physically would move a point of spacetime, AKA protons, neutrons, beta, delta, and another particle in the atom 5 times distant from where they were almost an instant before the expansion and at the speed of time, c. This near-instantaneous change in position of an entity embedded in a spacetime provides the enormous change in position, velocity, and hence acceleration for all of the subatomic particles involved in a fission or fusion, much beyond a chemical reaction, which does not change actual spacetime, only "bonds". And this instantaneous change in position was a result of the instantaneous change of the local coordinate system. Initially, you are here, instantaneously you are there. GR would say that your metric changed locally, that is the way you measure the distance between points, "proper distances", in that immediate locale has changed instantaneously, and became extended.

So that's external to the atom fragments or products, what was the cost to the inside of the initial reactants? Inside atomic boundaries, the products were somehow left with less spacetime curvature, less energy or mass, because the reactions were "exothermic". The change in mass was just energy loss, $\Delta E/c2$, in the form of a decreasing spacetime curvature due to the decrease of atom radius and internal spacetime curvature.

To understand this concept from a fractal universe position, consider our Universe and how it is known to be expanding. The expansion is due to the expansion of spacetime itself. The major theories start with the universe expands at a constant rate. Scientists can calculate how far away the edge of the observable universe is today by knowing the speed of expansion by redshift and blue shift measurements. The universe is a minimum of 92 billion light-years. It could be bigger, but 92 billion is just the largest reference point that scientists have today. Analogously, when an atom boundary is breached, the "universe" surrounding the atom expands at light speed, a constant rate, changing the local coordinate system and metric position of adjacent entities instantly thereby giving them velocities on the order of light speed. And this is what is measured, particles traveling fast as light and trying very hard to slow down as they enter lessor metric spacetime.

To be sure, the transition factor is much greater than 5, as will be shown below, the transition factor is related to the location of spacetime origin inside the atom 3-sphere, specifically the subatomic's spacetime distance from the atomic center, and scaled by the reciprocal of the distance squared to the extended spacetime location. For volumes, space, as a geometric transition occurs from one space to another, a volumetric change is mathematically generated, instantly providing all objects with a change in position from where they were before and after the spatial expansion. This change in position occurs in an instant in time, rate C, and is what we measure as fragment acceleration. The result of the spatial curvature expansion or contraction is what provides the fragments kinetic energies from atomic fission or fusion. The differential in spacetime curvature from the initial and final states is the "mass defect" that we measure. Of course, the result is tremendous accelerations of fragments and EM, and hence "kinetic energy."

On a parting note, it is interesting that only the most abundant element hydrogen-1 atom nucleus consists solely of protons. Every other bound nucleus atom has at least one neutron. The reason given is that the nuclear force doesn't allow all-proton or all-neutron systems to be bound, and protons provide more destabilizing force because of Coulomb charge repulsion. For low masses, the most stable configurations of nucleons tend to live near the region where the proton number equals the neutron number. But at higher masses, Coulomb repulsion becomes important and stability shifts toward larger numbers of neutron ratios, N/Z.

The current theory is that protons and neutrons have their separate shell structures in nuclei. If we start filling neutron shells while leaving proton shells vacant then Beta-decay becomes "energetically favorable" so that lower-lying proton shells can fill instead. Eventually, the neutron shells "fill" above the neutron separation energy, and any added neutrons immediately "back out".

This is easily explained via nucleons with similar spacetimes but different harmonic structures, harmonics on nucleon boundaries which can interact by exchanging spacetime curvature and emitting spacetime residual from the discrete spacetime required for harmonic shifting up or down of certain modes, modes that manifest as gluons, or other subatomic particles. These would be other packaged forms of ST yet.

How does this all apply to Warp Drive?

If the Chronature model is accurate, then we should be able to develop a current science fiction called a warp drive engine. Of course, we can follow the spacetime trail warped by a dense heavenly body as it warps spacetime, but this is not a practical method of space travel. There must be other ways to physically warp spacetime. In practical fact, we are generating spacetime curvature from transferring atomic and subatomic spacetime into our macro spacetime already through fission, fusion, matter-antimatter reactions. This is currently known in fission or fusion reactions as "mass defect" or "missing" spacetime. "Missing" because summing the mass of the product's and subtracting the mass of the reactants, leaves a residual called "mass defect". There is nothing defective about this mass except that the arithmetic doesn't add up. Mass defect in our model is the residual spacetime curvature that must translate from within the atom boundary after a breach to spacetime geometry, giving an instantaneous impulse seen as a spatial expansion, explosion, just outside the breached atom due to the transformation in geometry to which the residual spacetime must conform.

In the Chronature model, this mass defect is merely transformed spacetime curvature, which if captured, amplified, and directed can become the source of spacetime warp, and when directional, a warp drive engine. What needs engineering is a way to "collapse" spacetime curvature, take it from a source fission/fusion/matter-antimatter reaction generating "mass defect" spacetime. We then channel the newly created spacetime curvature just behind our present position, the spatial "expansion" immediately increments our relative motion, delta-v, at light speed time changes, delta t, accelerating, dv/dt, our position forward as if it were a fission product fragment following nuclear fission. The spacetime manifold "expansion" quantification process is articulated and cited above by non-Euclidean geometer Kulczycki with the spacetime transformation from the elliptical non-Euclidean, pre-atom breach, to hyperbolic, post atom breach, non-Euclidean Hyperbolic geometry, and then to plain flat Euclidean geometry.

When does colliding and splitting the same atom take less energy and momentum, not more?

Let's take another mystery from a real-world example in Nuclear Reactor design. Most power reactors are "thermal" reactors, fueled by nuclear fission from "thermal" neutrons. That is neutrons that are moving very slowly produce fission with much greater ease, than in comparison to the speed of light, or even "fast" neutrons, which subsequently provide the fission energy in Fast Breeder Reactors. Thermal neutrons have a much higher probability, "cross-section", of splitting a Uranium atom than a neutron moving very fast with much larger momentum, "fast neutron". This is counter-intuitive because one would think that an object that is traveling faster and hence having more momentum and energy would more likely split another object than if it were traveling much slower. This is the opposite with nuclear fission, neutrons traveling very slow, are much more likely to split an atom, than neutrons traveling very very fast. That is why thermal reactors dominate reactor design in the world of power reactors. These reactors are designed to slow "fast" neutrons produced in fission so that they can split another Uranium atom and create heat energy for power. Thus to break an atom apart one must hit it with a slow stick. The harder the neutron strikes a target atom the less the chance of a break or fission.

     The nuclear jargon for the measure of slowing fast neutrons down is in "lethargy" units. But even lethargy is not physically explainable because neutrons are "neutral" and hence do not attract or repel from sources to be slowed or captured. This does not stop physicists from measuring lethargy and knowing other practical characteristics that are useful in nuclear reactor design. However, our current physical models of the atom are not helpful here, and empirical data is needed for engineering designs. But the lesson is not lost, despite the lack of usable physical models of the atom, physicists and engineers still make due, but maybe not as fast, efficient, or as effective as possible with lack of good available computational models of the atom and molecules.

6 – Spacetimes in Collision

Local Non-Euclidean 4Fold Interaction

Einstein's math professor Minkowski delivered a paper titled "Raum and Zeit" at a scientific meeting in 1909, the gist was about spacetime as a complete entity and foundation of reality. Minkowski said "the view of space and time which I wish to lay before you have sprung from the soil of experimental physics, and therein lies their strength ... Henceforth space by itself, and time by itself are doomed to fade away into mere shadows, and only a kind of union of the two will preserve an independent reality." Chronature remains faithful to this foundation, as we embark on the path of spacetime entities as "unions", each of which will comprise a form of time and three dimensions of orthogonal space, each of which acts independently inside itself but can be transformed to other spacetime entities.

The mathematics of topological manifolds has provided many characteristics and properties of manifolds with many interesting solutions. These are of interest to us here for their ability to model and predict the atom and subatomic boundary deformation in physical interactions; to study the fabric of the spacetime boundary between inside and outside of the atom. Although homeomorphism seems to have the virtual unlimited capability in deformation, the physical phenomena of fission and fusion teach us that the atomic boundary can be breached and reformed to other derived 4Folds. Perhaps 3-sphere manifolds have special properties, ones which when imbued with certain physical constraints, will yield models which will not only predict mass defect better but explain it in physical phenomena and geometry. In any case, we can deal with 4Folds as unions of space and time, with some modifications.

How do 3-Spheres, atomic manifolds, physically break apart at collision time?

I use the soap-bubble model as an analogy for starters. Scientists working at CERN made some correlations from S-Matrix collision giving the mathematical properties of the Euler beta function. This caused a stir because this function was found regularly in the data, and so ended up in the String Theory name for string harmonics. This caused String Theorists to look deeper into the collision process of the S-Matrix for clues to the actual structures and components in the Standard Model. Clues that perhaps subatomic particles should be treated as vibrating strings instead of points, reinforcing the idea that this String theory had some legs.

We proffer another theory on these enormously important CERN S-Matrix particle interaction data regarding collisions. The Euler beta function can be described as a function of Gamma functions, Γ.

Although Gamma functions are prevalent In many areas, Gamma functions would be a natural underlying mathematical data construct to be obtained if atoms and subatomic particles were 3-spheres, which upon breaking up would form smaller 3-spheres, much like large bubbles breaching and forming smaller bubbles. This would take us quickly into the bubble surface tension physics the surface area for reformulation of the smaller bubbles. This would hence be directly analogous to spherical spacetime curvature breaches forming fragments, themselves following a spherical configuration as the parent(s). The formula for a spherical surface in n dimensions is given by:

$$S_n = r^{n-1} \times 2\pi^{n/2} / \Gamma(n/2)$$

Please do not the presence of the Gamma function. Imagine play soap-bubbles breaking up, some breakup but some breakups form into smaller bubbles. The surface area of the new soap bubbles is related to the area of the original soap bubble as a function of Gamma, Γ, as the new soap bubbles were formed from the original soap bubble surface area. By analogy, the atomic membrane will at some stress-energy collapse and form smaller 3-sphere membranes. These will form 3-spheres with areas complying with the formula above having the Gamma function. For a three dimensional sphere this will become the formula that we all learned as:

$$S_3 = r^2 \times 2\pi^{3/2} / \Gamma(3/2) = 4\pi r^2$$

For the 4 dimensional spheres, hypersphere, surface area, becomes

$$S_4 = 2\pi^2 \times r^3 / \Gamma(2)$$

In atomic membrane breach and reformation activity, one could expect that this would produce data with the underlying Gamma function, detected in atomic collision reformating experiments yielding vibrating string harmonic data. Imagine the process of a large soap bubble breaking up and forming smaller bubbles.

Fission is the best example of breached matter. What is called the "mass defect" is the matter that is converted to "energy" from the fission products themselves having less mass than the parent atom. From a standpoint of Chrono spatial dynamics conservation of spacetime curvature is the primary and governing law in such events. It is the residual, parent mass minus the daughter summed masses, equals converting spacetime curvature transition that "gives off the energy" produced in fission.

Upon a collision, how does nature physically transform one spacetime into another?

Of course to understand how the spacetime transformation occurs physically would give us knowledge about how we can more efficiently manage the transformation for our more efficient energy applications. Non-Euclidean Geometry becomes handy again, revealing the beauty of the physical from the equations if we can just infer them from the measurement that we make corrections. Here the mathematics of Mobius Transformations and Inversion comes to the rescue for colliding 4-folds of a very specific type. 4D spacetime topologies with their inherent and observed morphological properties aid us in graphically understanding the transformation from one geometry or configuration to another.

Physical transformation depictions mathematically can be modeled by complex inversion, whereby spacetime curvature inside of an atom's, spacetime, can become through the process of inversion, another volume of spacetime curvature outside of the atom's unit circle boundary.

For example, under geometrical inversion of a circle, on a complex plane, any given point or object with radius r inside of a circle with center q, inverts the object to a distance equal to the reciprocal of r, 1/r, translating outside of the circle to a position with an angle inverse to it its position angle inside the circle:

| Atomic STime | Macro STime |
| Coordinate system | Coordinate system |

$$z = re^{i\theta} \quad \rightarrow \quad 1/(re^{i\theta}) = (1/r)e^{-i\theta}$$

This equation for 2D can be extended to the inversion of a sphere. The position from an atom's center q, conversion of a point from an alternate spacetime is the reciprocal of the radius that a particular transforming piece of curvature inside the atom has from the center q. The angle of the transformed curvature piece's transformed position to the inside of the atom is twice the bisector normal angle as shown in the figure below.

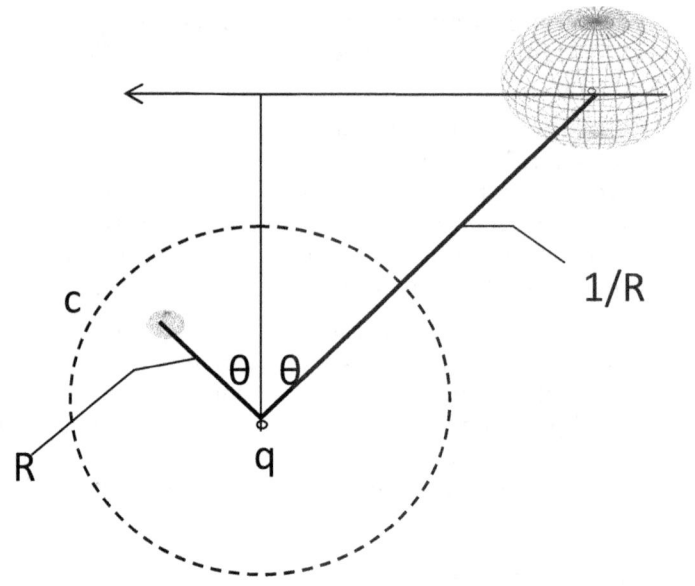

Such a transformation will manifest spectral lines, the transformation is from an atom's harmonic curvatures or "electron orbits" to EM, another spacetime. The unit circle C plays a surprisingly special role for this mapping since the inversion transforms the interior and exterior of C while each point in region C remains fixed. But this only gives us the spatial or radial coordinates in ST. This would also indicate that nuclear "subatomic" objects of spacetime curvature inside the atom boundary would transform to objects of curvature outside of the atom's boundary at a distance inverse to their inside position from the atom center and at an angle reflected but equal angle. Moreover, there is a strict analogy between inversion in a circle and reflection in a line. That is if a line L does not pass through the center q of a circle K, then inversion[35] of K maps L to a circle that passes through q. Furthermore, "if a circle C does not pass through the center q of circle K, then inversion in K maps C to another circle not passing through center q."

But here is an important clue from the mathematical to the physical. The internal object's distance, R, from within the atom center, transforms to a distance of $1/R$. Therefore the closer to the center of the atom, the farther the transformation occurs to outside the atom. Keep in mind that 3D surfaces will have the R^2 curvature factor, or the reciprocal, $1/R^2$
in the transformation of curvature, aka energy, from one spacetime to another. That would be in space-space and spacetime surface curvature.

In addition and as expressed above, parity is also a property of spherical harmonics. The harmonics are either even or odd with respect to inversion about the origin. Inversion is represented by the operator $P\psi(\mathbf{r}) = \psi(-\mathbf{r})$. Then, as can be seen in many ways (perhaps most simply from the Herglotz generating function), with $\mathbf{r}$ being a unit vector,

$$Y_l^m(-\mathbf{r}) = (-1)^l Y_l^m(\mathbf{r})$$

In terms of the spherical angles, parity transforms a point with coordinates $\{\theta,\phi\}$ to $\{\pi-\theta, \pi+\phi\}$. The statement of the parity of spherical harmonics is then:

$$Y_l^m(\theta,\phi) \to$$
$$(-1)^l Y_l^m(\pi-\theta, \pi+\phi) = (-1)^l Y_l^m(\theta,\phi)$$

---

[35] *Visual Complex Analysis* by Tristan Needham pgs 128-129

Strange as it may come, simple angle of incidence equals the angle of reflection of photons off of surfaces has always been very difficult to explain with pre-existing models and under QED. Feynman[36] says it most honestly, and after he explains the theory. "The situation today is, we haven't got a good model to explain partial reflection by two services; we just calculate the probability that a particular photomultiplier will be hit by a photon reflects from a sheet of glass." And yet under geometric inversion shown above, angle of incidence and angle of reflection are simply transformations in geometry, motions from one space to another. No probability amplitudes are needed and "counting beans" becomes from a physical view, understandable.

IF QM is "incomplete" how is it that Quantum Entanglement has been proven to work as advertised?

The counterintuitive predictions of QM about strongly correlated systems with the same birth were formulated in a thought experiment by Einstein, Podolsky, Rosen and known as the EPR paradox. This was to show QM was incomplete, leading to "spooky" results if proven physically. An "entangled" system is defined to be one whose quantum state cannot be factored as a product of states of its local individual particles. If entangled, one pair constituent entities cannot be fully described without considering the other pair constituent. The seeming

---

[36] *The Strange Theory of Light and Matter*, Richard P. Feynman p24

paradox is that a measurement made on either of the paired particles apparently "collapses" the state of the entire entangled system into one state instantaneously before any information about the measurement could have reached the other.

Remember that QM is strictly a probability model already and mapping it to anything physical cannot be done without a ton of faith, conjecture, and statistics. QMers were challenged by Einstein and the company to explain the inconsistency in the EPR paradox in particle quantum states and so had little choice but to prove out the paradox or admit that QM theory was desperately awry. Over the years QMers sought to create experiments to show that the paradox represented a real physical phenomenon and that QM was indeed a valid model. Since it's not possible to prove the paradox for any single physical pair, not individual pairs but groups of pairs of particles were used. This Quantum Entanglement is a physical phenomenon that occurs when groups of particle pairs are generated or interact in ways such that the quantum state of each particle cannot be described independently—instead, a quantum state may be given for the system as a whole. Measurement of groups of physical properties such as position, momentum, spin, polarization, etc, performed on "entangled" pair groups of particles are found to be "appropriately" correlated. In all of these experiments, causation is never proved, only correlation. Furthermore, individual particle pairs are never used, only groups of pairs and so more statistics are used. Thus nothing can be conclusive of the EPR paradox experiments except that QM is an incomplete theory at best having a tough time explaining common sense physical phenomena.

But there is something important about the EPR paradox that screams for an explanation since in some QM Entanglement experiments, Delft, apparently, successfully shows correlations between groups of paired particle states. And so we are compelled to provide a brief answer from a simple mechanical model of pair production.

To do this by analogy we first reach out to a physical phenomenon well known in Fluid Mechanics called Vortex Shedding, Strouhal affect. The vortex shedding phenomena in fluid flow around a rigid body produces pairs of opposite circulating vortexes, let's call it "pair production". This is completely analogous to ripping off bonds or "entanglements" between atoms or photons or subatomic entities in a field barrier. The entities or pairs produced will seek to minimize the Lagrangian of the spacetime curvature. And there you have "entangled" oppositely formed pairs of entities, but with a physical understanding of exactly what is happening.

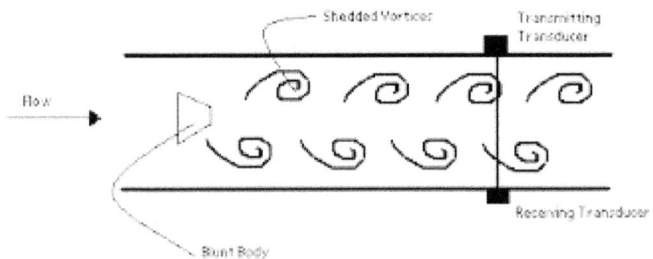

Not unlike entanglement pair production, vortex pairs are created as fluid flows past the blunt-body shown in the center of the flow stream, where one of the pair members is formed first to the left or right with a directional circularity property followed by the other member constituent of the pair formed then to the right or left respectively with the opposite circularity state. The transmitting transducer and the receiving transducer need only measure one of the pair vortex members to be able to "determine" the state of the other and instantaneously, that is faster than c! So here we have quantum entanglement by analogy to the physical phenomena in fluid mechanics vortex shedding. Nothing spooky here, except for perhaps a complete abandonment of scientific thought and common sense.

The fluid stream is shown in a confined channel, but if one constituent vortex moved away, say to Rome, and the other little vortex stayed home, their relationship would not change. Measuring one pair constituent would instantly determine or "collapse" the state of the other pair constituent and not because of some mysterious spooky action, but because the pair relationship was determined at creation time, not measurement time. I find it hard to believe anybody would take that any other way since the experiment begins at pair production time and state.

Let's quantize this down to the quantum order realm. From Chronature we know that all particles and entities are spacetime curvature manifested in different spacetime curvature configurations or attributes of states. The atom and some subatomic particles are 3-Sphere curvature spacetime vortices, and these spheres have many types of symmetry attributes. Pair production would naturally provide that spacetime curvature was conserved, our Primary Conservation Law, and hence pair production would naturally produce paired entity states that are polar, symmetrical, circular, temporal, valence, or sign opposites. Measuring the state of one pair partner would obviously "instantaneously determine" the state of the opposite pair member. It's called a "duh". The difference in this explanation and any that I have seen on Quantum Entanglement is that it's a physical phenomenon depending on how the single pair "entangled" or was created, and after pair creation, time for determining the other pair member through measurement is irrelevant because it states is already set, its the opposite of the measured pair partner. Hence quantum entanglement has little to do with QM, and much to do with actual and real physical phenomena of opposing symmetry in pair production. EPR paradox throws a cog into the wheels of QMers forcing them into a less arbitrary way to choose an entity's state.

But there is a profound aspect to the EPR paradox, and that exists because of the Conservation of Spacetime Curvature Law. With all the matter created at the big bang and still being created, it seems intuitive that for the spacetime conservation to hold, all matter must have opposite and "annihilating" spacetime entities. Some have postulated that this would come in the form of anti-matter. Therefore pair production and annihilation would ensure that spacetime is conserved. For every entity with Yin spacetime curvature, there would be an equal and opposite Yang spacetime curvature in some dimension or sense. Hence at creation time, an entity may have a pair or more. There does not appear to be sufficient anti-matter to establish zero equilibrium but perhaps Black Holes exist as the annihilating reservoir for the other pair members or perhaps there is something we cannot "see" like Dark Matter with the opposite but undetectable by or transparent to EM character. But the physical truth may be just as simple as pair production of subatomic and atomic entities, whereby the pairs spacetime at production is conserved, by the opposite curvatures affected by entity pairs with opposite states, ie opposite spacetime curvatures.

QMers assertions to quantum entanglement proofs as "hidden variables", or "local hidden variable" or "local realism" and others are well known. "Entangling electrons" and then trapping them in separate crystals for measurement does not prove much more than the fluidic shedding analog "entangling" vortexes at creation time. Somehow scientific explanations using a bad model do not overcome realism or reality.

Scalability of a Unified Field Theory

Gravity waves are difficult to catch. But even in 1916, Einstein considered their existence likely and demonstrated mathematically that gravitational waves should arise when you replace a spherical mass with a rotating dumbbell of the same mass which, due to its geometry, will generate dynamic ebb and flow effects on spacetime. A 3D visualization of gravitational waves produced by two orbiting black holes, courtesy of Henze at NASA[37] is shown directly below.

Gravitational waves are ripples in the curvature of spacetime that are caused by collisions of heavy and compact objects like black holes and neutron stars. Where especially large masses are moving wrt and proximate to each other, gravity waves of significant magnitude are generated, waves of spacetime curvature. But for our planets and sun, the magnitudes would be too small to detect with anything that we have today. Many such planetary masses moving will emit gravitational waves, but they will be so incredibly weak that they are by us for all intents and purposes undetectable. It takes very massive, compact objects to produce already tiny strains that can be measured.

Although orders and orders of magnitude difference it is noteworthy to compare the picture above for Black Hole structural similarity to the

---

[37] http://www.universetoday.com/83759/astronomy-without-a-telescope-unreasonable-effectiveness/

Chronature model and spacetime harmonic modeshapes shown in previous chapters and as modeled by the vibrating sphere, the S and P levels, in particular, vibrating sphere solutions. The superposition of the vibrating sphere harmonics and atomic orbital minus their dimensional differences shows that the mathematics for galactic spacetime curvature is similar and analogous to atomic levels, with both radiating spherical harmonics. This is another strong argument for a fractal spacetime universe, where spacetimes have smaller spacetimes are embedded and with similar characteristics but scaling up or down, where 4D manifold dimensions are scaled from atomic units and up to astronomical units. The Swartztchild radius physics may yet play a part in atomic physics for understanding the bond interactions between atomic spheres.

The Penrose diagram helps describe the spacetime coordinate geometry for Black holes, and a similar technique may be useful in application to the atom spacetime. However, the atom boundary curvature on a sphere is the same for all points on that sphere, similar to the Black Hole, differing by the radius of the sphere. Moreover, the curvature of the harmonic mode shapes will be different at most points. But where the lobes are symmetric, there will be points of equal curvature on similar lobe positions. And we will not need the EFE to calculate the curvatures there since we will have closed-form solutions to the shapes.

What is the fundamental difference between the modern physics Standard Model and the Chronature Theory?

Modern physics models matter at the most fundamental level as fields and particles are affected by these various fields. A field is an abstraction, a non-physical entity representing an acceleration vector, and where matter is just a set of particles and wave packets that obey the different fields that are labeled with names near the forces that comprise their measurements, all of this without any interpretation or understanding of what they represent physically.

Modern physics describes nature in terms of abstractions labeled fields. Each field has a description as the set of defined and labeled particles. A "force" between two particles can be described either as the action of a field from a particle, or a force "carrier"

The Standard Model contains a set of particles, each of which is an excitation of a particular field. These are, for example, Gluons, excitations of the strong gauge field, Photons, W bosons, and Z bosons, excitations of the electroweak gauge fields. Higgs bosons, excitations of one component of the Higgs field, which theoretically gives mass to fundamental particles, unknown as to how that is done or what "mass" really is, and several types of fermions, which are themselves described as excitations of fermionic fields.

In the Chronature model, all fields are simply spacetime curvature producing acceleration fields which are dependent on embedded spacetimes, subatomics, warping around disparate spacetime to produce the acceleration fields. The subatomics embedded spacetime is just another kind of spacetime and different from the atom spacetime but still having three spatial dimensions of non-Euclidean geometry and one-time dimension, perhaps a negative frequency-time to distinguish that from the atomic spacetime. These spacetimes also comprise 3-sphere manifolds which have harmonic frequencies and modeshapes, which determine their characteristics and fields through their spacetime boundary curvature.

7 – The Intersection of Light and Matter

What is a photon? Optical scientists and engineers fall back on the Maxwell model shown below. The present modern physics and QM models have not progressed much past, Maxwell. Einstein introduced the concept of "indivisible quanta" from his Photoelectric data 20 years before QM came on the scene and this no doubt helped propel QM through and much debate. But the wave-particle duality hypothesis remains unresolved. This huge deficiency of a good physical model has precluded prediction of light-matter interaction and experimentation trial and error are the dominant method for gaining much in understanding EM. The very structure of a photon remains unknown. In Richards Feynman's theory of Quantum Electro Dynamics[38] – he admits to what the physicists do to get the right answer, which does not explain photon transport, but is instead a probabilistic argument to "get the right answer."

> "I have chosen this calculation as our first example of the method provided by the theory of quantum electrodynamics. I am going to show how we count the beans – what the physicists do to get the right answer. I am

---

[38] *The Strange Theory of Light and Matter* , Richard P Feynman,

not going to explain how the photons actually decide where to bounce back or go through; that is not known. Probably the question has no meaning. I will only show you how to calculate the correct probability that light will be reflected from glass of the given thickness because that's the only thing physicists know how to do! What we do to get the right answer to this problem is .."

In another phenomenon, Rayleigh elastic scattering of electromagnetic radiation on impinged matter particles with a radius much smaller than the wavelength of the impinging EM radiation. From the standpoint of Chronature models, this would be large diameter sphere trains of Elliptical geometry space impinging on smaller entities of spacetime comprised of elliptical non-Euclidean geometry and frequency-time. The interaction or scattering between these does not change the state of atomic entities and hence it is a parametric process. The impinged spacetime entities may be individual atoms or molecules. This interaction can occur when light, sphere trains, travel through transparent solids and liquids but is most prominently seen in free spacetime atoms or gases.

The traditional teaching is that Rayleigh scattering results from the "electric polarizability" of the particles. The oscillating "electric field" of a light wave acts on the charges within a particle, causing them to move at the same frequency. The particle, therefore, becomes a small "radiating dipole" whose radiation we see as scattered photons.

Thus the current state of the leading theory of QED is that EM and its physical interaction with matter are not well understood from a physical model and that the physical model "has no meaning". The theories retreat to a probability argument that a particular event would happen, probabilistically and without a model for how it does happen physically. But spacetime geometry can be configured as EM, light is the most popular, actually does have a physical structure and we model that physical structure as spacetime curvature to explain the unexplainable of how and why light propagates in straight lines.

The classic Maxwell model structure of EM is shown in the figure below, complete with "E-Field" and "B-Field".

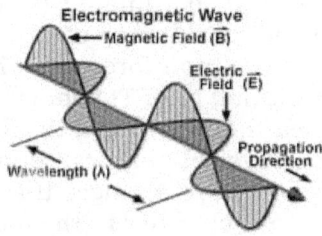

The wavelength and frequency of EM are easily discernable in this representation of the Maxwell model, although wave amplitude is always shown but never referenced. The use of peak amplitude is simple and unambiguous for symmetric periodic wave-like EM, but 3D physical reality is missing. How does this model depict physical reality? What does that "E" field and "B" field represent in physical reality? While the diagram looks like a physical model, it is only a schematic and for that reason, we cannot know the physical reality from EM in this model. A more 3D model may look something like this for one period or wavelength:

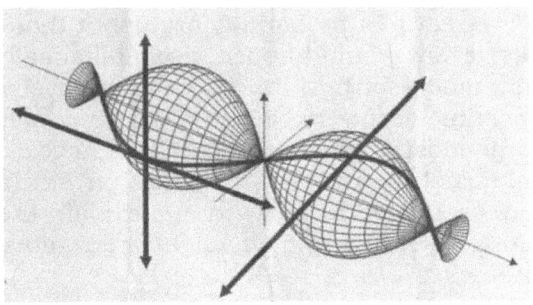

where the coordinate system moves along the travel axis as it rotates along the surface. Thus a straight line show the local coordinate system axes in bold rotates freely about the propagation axis. The geodesics, paths of non-forces are thus laid out with the moving rotating grid. The longitudinal axis is typically called the E field and the latitudinal axis is the B field. A physical manifestation of EM spacetime curvature would show changes that appear to work orthogonally to each other in the form of Lorentz force. Spoiler alert, the "E" and "B" fields are in physical reality moving magnetic fields with the repulsive and attractive one-sided surface, having North and South poles, pushing and pulling the photon along a geodesic curvature path, separated by a topologically twisted time neck morphed Klein bottle. But this geometrical of a rotating twisting coordinate system of spacetime is sufficient to explain what is happening in physical reality.

Moreover, in one model for light interaction with matter is soliton entities of spacetime curvature structures impinging on a vibrating matrix of spacetime curvature lattice structure bound nodules, atoms, effecting changes from absorption on some natural harmonic vibrations and not others. One of the most important variables of a soliton is its amplitude. The amplitude and hence impinging curvature becomes the essential quantity in the understanding of the physical interaction between light and matter.

The amplitude of the EM wave, from the classic Maxwell model shown above, is sometimes incorrectly redefined as the "intensity". The amplitude of EM curvature structure photon is ignored. Yet the amplitude is most closely related to curvature emanating field from the photon centerline and hence our interest here.

A recent picture of light itself, photons, was captured by a new kind of "camera". Scientists have photographed light behaving simultaneously as both a particle and a wave. A picture of the dual nature and structure of light was captured by scientists at École Polytechnique fédérale de Lausanne (EPFL), Switzerland, due to an unorthodox imaging technique. The scientists generated the image with electrons, making use of EPFL's ultrafast energy-filtered transmission electron microscope. A picture is shown below.

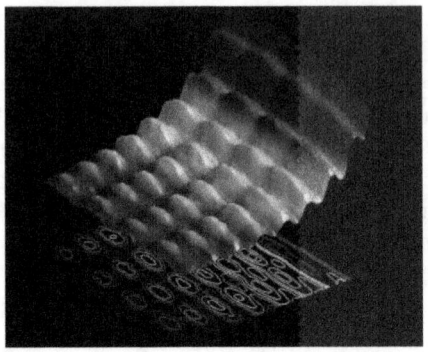

One can see that the structure is parallel oval-like kinked trains and the boundaries are not smooth, which is probably due to the lack of granularity in the imaging technique. However, it does show that the photons travel in a parallel spherical-like train car structure giving it the particle and wave component character. Moreover, even a solitary lone photon can explain "single-photon inference" as a single photon will have multiple "cars" in a "train" of F-Spheres, having the alternating "E-Field" and "B-Field" components in each "car".

Outdated as it is, still, the best that we have on light-matter interaction is from Feynman's work. Feynman's QED provides an answer, albeit a calculated one using "probability amplitudes", not a physical model. Since QED does not provide a physical model, the arguments that lead to the correct numerical, "counting the beans", answer are probability-based. Instead of a light ray hitting a mirror and reflecting at the same angle, as we have been taught, Feynman's QED argument [39] is that a "light ray has equal physical probability of reflecting from all points of the mirror regardless of angles." Thus a light ray's incidence equals reflection rule is conjured by working backward, knowing the right answer, and constructing an argument that starts from probability and ends in the correct answer, the rule of incidence equals reflection. This argument, brilliant and amazing as Feynman was, never felt right and so I give you Chronature theory, taking from the physical realm and constructing a composite spacetime model of EM, a model based on mathematical rules of Mobius Transforms and Complex Inversion that determine the physical outcome that we see.

Looking back on the classical figure of an EM wave two important components are missing, 1) the rotation of the entire wave train is not shown and 2) the amplitude of the wave is completely ignored. The rotation about the travel axis is sometimes referred to as circularity, right-hand or left-hand circular. As it turns out, this is a very important aspect of EM. But if we visualize the rotation, it would look like a train of sphere cars attached at the axis of travel or time. I have included some figures that show the geometrical boundary of EM waves so that reflection and scattering can be understood better.

How is a massless photon affected by gravity?

---

[39] QED – The Strange Theory of Light and Matter, Richard P. Feynman, pg.40, "The quantum view of the world says that light has an equal amplitude to reflect from every part of the mirror"

Photons are affected by gravity mostly because they are following the curvature of the spacetime warped by a large mass.

In the current models, photons can have energy per particle without mass because they obtain relativistic mass, "rest mass", which is not mass in the classic sense. And after all, those speedy electrons need rest at times. Their energy, $E = pc$, is their momentum multiplied by the speed of light. This kind of explanation is due to the best photon models that physics offers today.

In the Chronature model, photons are transferred moving spacetime curvature into atom harmonics at atom harmonic resonances and hence they gain "energy" from curvature transition from moving wave curvature to standing wave curvature. Inside of an atom spacetime, time is oscillatory or frequency, and once transferred the frequency-time, $1/c$, becomes c or speed of transformed spacetime curvature entity a velocity, spacetime oscillations, of the converted monotonic time. It is well known that photons travel at time speed, also light speed, c, but traditional models hold that their "momentum" is an empirical measurement, not something physical, and brought to us from the De Broglie's relation:

$$p = h/\lambda$$

The equation for energy E is classically

$$E^2 = p^2c^2 + m^2c^4$$

These may not be physically useful quantities because there are no physical models to expound and explain the parameters and numbers in the equations. Thus currently energy and mass can be positive or negative or imaginary. Tachyons are particles with an imaginary mass, but before

Chronature there are no physical models to help us understand what is physically manifesting, only probabilities of potential states.

Why do the new solid-state Wakefield and Plasma chip-size particle accelerators work?

Particle accelerators are the big tool scientists currently use to study the structure of matter and sub-matter. Most recently a solid-state accelerator technology has emerged in the form of Wakefield accelerators. An explanation of Wakefield chip-size accelerators, in is as follows. Wakefield accelerator phase-locks a particle bunch on a photon wave and this loaded space-charge wave accelerates them to higher velocities while retaining the particle bunch properties. Currently, plasma wakes are excited by appropriately shaped laser pulses or electron bunches. Plasma electrons are driven out and away from the center of wake by a non-linear force component or the electrostatic fields from the exciting fields, electron or photon. The plasma ions are too massive to move significantly and are relatively stationary at the time scales of plasma electron response to the exciting fields. As the exciting fields pulsed through the plasma, the plasma electrons experience a massive attractive force back to the center of the wake by the positive plasma ions that have remained positioned there, as they were originally in the unexcited plasma. This forms a full wake of an extremely high longitudinal accelerating and transverse focusing electric field. In short, each pulse sends a cluster of photons into another cluster of electrons accelerating them forward. Even as I write this I'm thinking "what a load!", the wave shown in the diagram is EM, that is it not solid or liquid, it's just a model representation of a non-physical model or picture of a wave. From which a small sphere is bouncing in this non-physical trough just behind, in the wake of, a bigger sphere that is laying the wave down like a road pavement machine.

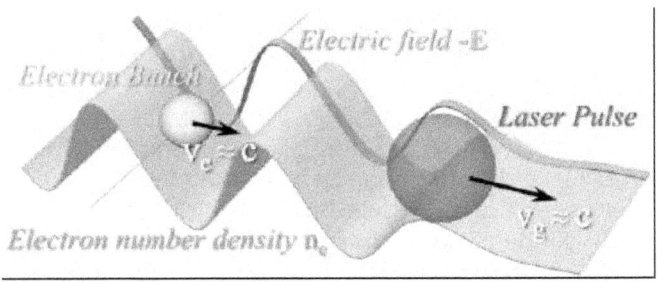

How does the Chronature model explain a photon accelerated electron?

Acceleration by a photon from Spacetime curvature is shown in the below diagram. On a Minkowski spacetime graph, a flat spacetime path left side is shown adjacent to the system or co-ordinate Time axis. That is the spacetime curve in a flat or empty space. Towards the right of flat spacetime is an electromagnetic wave worldline.

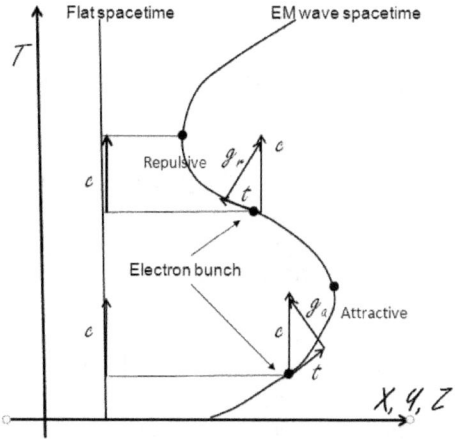

At local points on the EM, the worldline path is shown, as Wheeler named it, the momenergy vectors, with components of spacetime curvature aka acceleration, repulsive and attractive, as well as proper time τ. The accelerations, labeled repulsive and attractive, go back and forth and cancel each other in the spatial direction but add in the time direction. That is to say that spacetime curvature alone easily explains the acceleration of an electron by photon transport in a Plasma or Wakefield accelerator. The important components of the Wakefield accelerator could be obtained by inspection of the chart, ie. the dimension of the EM wave in space, the frequency of the EM fluctuation in time, and the physical dimension and inertial qualities of the particle. Here the electron is shown as negligible in mass, therefore inertial effects minimum, but the field strength of the electron, the direction of the field in space, the pulse frequency, and the laser frequency are all physical parameters the would affect the accelerator chip dimensions.

How does Chronature explain travel backward in time?

Feynman proposed that a positron, anti-electron, was an electron traveling backward in time. The figure below shows an electron-positron pair annihilation in the Feynman diagram (b) below. Feynman states[40]

> "that the electron emits a photon, then travels backward in time to absorb a photon, and then proceeds forwards in time again. The path of such a backward-moving electron can be so long as to appear real in the actual physical experiment in the laboratory."

Although this is not physically explainable Feynman hedged with the probabilistic theory, falling back to his earlier admission that although QED is the "crown jewel" of physics, yet he didn't understand how the light-matter interaction worked in physical reality and claimed, "nobody does."

---

[40] Richard A. Feynman, *QED – The Strange Theory of Light and Matter*, pg 9, pg 99

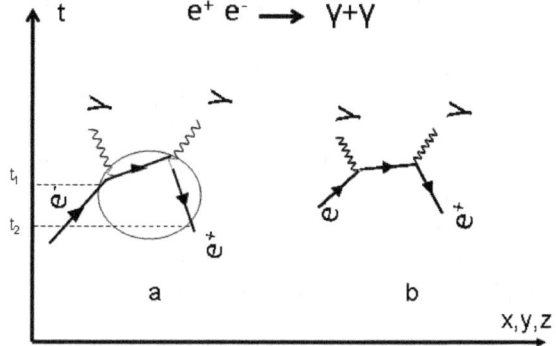

The positron is a very real phenomenon, and well applied in the medical field as in PET scan devices. This is yet another example of how scientists and engineers can use nature without a complete understanding of its nature. Empirical and statistical methods are used where direct models are unavailable, but those results are severely limited and narrow as they generally employ random variables to very specifically limited conditions and scope, only because deterministic models are unavailable.

However, the physical light-matter interaction mystery is simply explained in Chronature and shown in the figure below in the frequency-time spacetime model inside the atom. The right-hand side (b) shows the typical Feynman diagram of the electron-positron annihilation forming the gamma rays. The positron is indeed shown traveling backward in time. On the left-hand side figure and atom spacetime is plotted on the classical Minkowski axis (a) depicts the interaction and formation inside the atom, where time takes on the spacetime that it enters, frequency-time, and the "positron" is formed there. Indeed the time of entry outside of the atomic spacetime t1, where the frequency-time was governing, a positrons polarity is reversed and its exit time t2 appears to be rotated to before the time that it entered the atom's spacetime boundary. Hence the Chronature model lends it to easily explain how the electrons travel backward in time, a positron by definition; acquiring a frequency-time while the electron traveling forward in time remains in its monotonic increasing proper time, and is thus traveling "backward". This is different from negative time, which was explained earlier from the standpoint of a macro time curvature in the Minkowski spacetime axis and the newly forming universe front. But an important concept to note in this matter-anti-matter interaction model here is that the positron is not a new particle, it's the same particle, electron going backward in time, interacting differently with matter, causing us to think in our philosophy that it's a new particle when in fact it is an electron going backward in time, t_2.

Much activity has been concentrated on attempting to understand positrons and their "atomic" model see below figure, called Positronium.

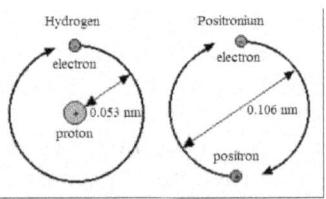

As usual, theories and experiments abound and all over the map with data. Positronium is currently modeled as a "Hydrogen-like bound state of an electron and a positron". The 2 different "states" live from 0.125 to 142 nanoseconds. This should tell you that these models are too primitive to describe Positronium, as an atom with atomic states, but better as in Chronature, a transition of spacetime from one form to another as shown in the modified Feynman diagram above. Where the positron is an electron "going back in time" having entered another spacetime, a CTC – Continuous Time Curve spacetime of Gödel, generating a spacetime boundary harmonic transmitted to a spacetime system time before entry. Physically this would be a topological Klien bottle one-sided surface model transition exhibiting an atomic boundary time "inside" curvature being turned into an "outside" curvature showing a charge inversion from negative to positive.

Antimatter

Antimatter is material composed of antiparticles; which ostensibly have the same mass as particles of ordinary matter but have opposite charge and perhaps other particle properties. Experimentally shown collisions between particles and antiparticles lead to the annihilation of both, giving rise to variable proportions of intense but identifiable signature photons and less massive particle-antiparticle pairs.

From the low number of antiparticles produced it appears from scarce data-limited conditions that antiparticles can bind with each other to form antimatter, just as ordinary particles bind to form "normal" matter. For example, a positron, the antiparticle of the electron, and an antiproton can form an antihydrogen atom. But this neither is the extent of human's ability to produce antimatter nor observed in nature. Studies of cosmic rays have identified both positrons and antiprotons, but this also seems to be the extent of antiparticles with a few exceptions.

The best theory to date comes from the Feynman–Stueckelberg interpretation which holds that antimatter and antiparticles are regular particles traveling backward in time due to up-spin decoherence. Positrons can be produced by radioactive $\beta+$ decay, both naturally and artificially. Antineutrinos are another kind of antiparticle created by natural radioactivity ($\beta-$ decay). Antiparticles are also produced in any environment with a sufficiently high temperature

An antimatter experiment was made at CERN for Antihydrogen. CERN's production capacity requires 100 billion years to produce 1 gram. The only other antimatter produced and at substantially smaller rates is Antihelium. We only bring them up here because there is much speculation but little really known about antimatter. And perhaps this is because we have no physical models of matter, logically leading to no real physical understanding of the phenomenon of anti-matter.

From the Chronature model, there must be a natural purely spacetime geometry accommodation for antimatter, a physical explanation for this much speculated upon phenomena. Our clues are that the antiparticle has the same mass as particles of ordinary matter but opposite charge and "complete annihilation" upon contact. We would use the Chronature model for charge physically, the asymmetry of a spacetime entity boundary creating an attraction force to another curvature preferentially to a particular particle boundary. But total reactive annihilation of matter is the big deal because it indicates the total conversion of small amounts of matter to large amounts of EM or "energy" This concept is used as the energy that powers the warp drive engine in science folklore and is probably the reason for the tremendous amount of interest that it receives.

But if our model is worth its weight, we must explain what is physically happening when matter meets anti-matter. One way to explain these physical phenomena of matter-antimatter"annihilation" or total "energy" in the physical Chronature model is to introduce a negative frequency-time dimension inside an atomic boundary, defining another and oppositely different spacetime. The harmonics on the atom boundary, electrons, will then be opposite in frequency, producing precisely opposite and countering spacetime curvatures. Just as a pulse of negative curvature perturbation meets a pulse of positive curvature perturbation where and when they meet the sum is flat, just like two opposite phase waves in passing. But if "matter" were to meet with its out of phase opposite, we have two spacetimes with opposite sign frequency-times, and out of phase modeshapes in spacetime. Upon meeting they would "flatten" time to zero, their time component dimension would transition to another time dimension and the matter, spacetime curvature, would transform to another type of spacetime, that of a traveling EM in the macro spacetime. This transition then appears in the form of conformance to the new spacetime.

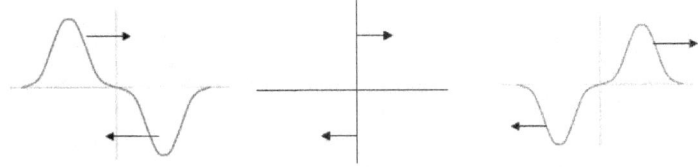

The figure above shows the opposite traveling wave interaction with regards to interaction within the same spacetime and is shown here only as a partial analogy, where wave curvatures meet. In two opposite frequency-time spacetimes, curvature must conserve dimensionality number of four, and transform into a 4fold matter-antimatter "annihilation" transformation into the macro spacetime completely. This means "electrons", which are standing wave harmonics on a spherical spacetime membrane, become traveling wave curvature entities taking on macro time dimension in macro or the embedding spacetime. The right-most figures would not exist in that spacetime form as it would necessarily be converted to another but equivalent spacetime once the frequency-time and negative frequency-time dimensions acted to conserve the interacting space curvatures.

One law to rule them all: SpaceTime Curvature Conservation

According to the discussion just above on antimatter, we begin here with the Conservation of Spacetime Curvature Law. The three basic current conservation laws as we know them logically emerge from the conservation of spacetime, which is that spacetime curvature from one kind of spacetime to another spacetime or kind of spacetime, must be conserved inconsistent mathematical fashion. This conserved is manifested in the three basic laws of conservation that we accept; conservation of energy, conservation of momentum, and conservation of matter within a closed system. These three laws of conservation are derivable from the basic Primary Conservation Law of spacetime curvature.

The basic conservation laws come from the classical Newtonian concepts and are preserved in the Einstein Stress-Energy-Momentum Tensor. This was done by Einstein so that his GR would collapse to the classical solutions measured at the non-relativistic conditions known to that point. But aside from that, only the geometry of spacetime needs to be satisfied and consistent, opening the door to having something even more fundamental responsible for the current conservation laws.

From a simple atomic spacetime perspective, this can be shown as follows. The curvature of a surface subtending a spacetime volume inside at radius R_1 is the total of the curvature over the area of the subtended volume; $(K/R_1^2)(4\pi R_1^2) = 4\pi K$. The stretched or extended curvature at another radius in the atomic 3-sphere at radius R_2 would be $K/R_2^2 \times 4\pi \times R_2^2 = 4\pi K$. Assuming the temporal dimension is analogously similar at both radii, this argues for the conservation of spacetime curvature in this simple slice of curvature. Since the geometries of atomic spacetime curvatures are unique to each atomic element and not necessarily perfectly spherical but undulating spherical wave membranes, this is a special case using average radii provided to show that such a thing comes about from the properties of the geometry of calculating spacetime curvatures.

From a physical model perspective between spacetimes, the physical topological properties of the photon and matter, Bosons vs. Fermions, become important in describing the morphing, Topological isomorphism, or transition from one spacetime configuration to another. This morphing transformation will involve typical topological functions including Reflexivity, Symmetry, and Transitivity, in homeomorphism, diffeomorphism, and isotropy, equivalence relations, between the physical structure of EM and the structure of the atom in their respective spacetimes. What sounds like a load of gobbledygook is borne out as the mathematics of topology which comes to the rescue in explaining the physical properties of 4D surface boundaries, volumes, and the interactions of different 4-folds.

Furthermore, the geometry of spacetime interactions becomes rubber-sheet geometry of topological spaces where straight lines and dimensions give way to contiguity. Valid geometry of translations and mappings including elongation, inflations, distortions, and or twists remain to describe the interactions and hence model physical reality. Under Chronature models, computer renderings of 4D manifold interaction calculations will supersede the world of the experimentalist. Hopefully, to provide precise answers to basic questions based on precise geometrically calculable physical models, for example, the how/when/where atomic and subatomic collisions occur.

Homotopy, a Topology mathematic concept, may be applied to understand physically what is going on in the EM-matter interaction. If one continuous function can be deformed into another, such a deformation is called a homotopy. For example, a torus is homotopic and isotopic to a coffee cup, one can be deformed into the other and vice-versa. One Chronature conjecture is that the atom's discrete harmonic modeshapes are topologically homotopic or isomorphic with corresponding EM radiation, mapped frequency EM ray. The curvature is deformable from one to the other. Please see our F-sphere, deformed from a Klein bottle, and with a rotating time dimension deformable to EM, having in monotonically increasing time dimension physically manifesting a series of F-sphere cars in a train.

Why do we have a Conservation Law of Spacetime Curvature?

Just announcing that Spacetime Curvature is conserved is not sufficient to explain the why. All through the application of conservation spacetime curvature will provide good physical answers and results, that doesn't explain the why either. The answer lies in symmetry. For the GR geometry literate, where we have a metric that remains the same, we have symmetry. A conserving metric, where the metric does not change, distance preserving mapping between points in the spacetime, an entity's travel along a curve where the metric does not change, will not feel any force. A geodesic is one such entity and hence any configuration of 4D manifold comprised of geodesics will be spacetime curvature conservative. For the math guys, this means that the Lie derivative of the metric is zero. This entails calculating something known as a covariant derivative.

What is the physical structure of a photon?

Kaluza's theory of unifying Gravity and EM was accomplished by adding another spatial dimension to the metric tensor and Einstein's Field Equations. Despite this mathematical breakthrough at that time, the physical manifestations and representations remained debatable. But transitions from one spacetime to another gives the structure of a traveling EM wave as a straight multiplication of a time vibrating standing wave in 3 spatial dimensions, producing a spherical harmonic duplicated and "connected" normal to its spherical surface in the direction as a rotating axis spherical wave train, see the bowling ball picture below. One can see here why physicists like looking at EM from the standpoint of Wavenumber, the spatial frequency showing the number of sphere trains per unit length.

The transition harmonic frequencies of spacetime inside the atom that comprises the photon train, apply a reverse inversion or Mobius to the outside atom spacetime and may have a squished spheroidal configuration to them while retaining the sphere rotational dynamics along the axis of travel. The Chronature model of a photon is that of a sphere train in the direction of travel, whereby each sphere in the train must transform from inside a 3-sphere where frequency-time confines curvature in modeshapes of characteristic harmonics, which are orthogonal, to the same atomic volume. The figure to the right of the sphere train attempts to show the effects of Lorentz contraction, a relativistic phenomenon that contracts the dimension of propagation at relativistic speeds. This contraction is what gives the magnetic field effect, where straight lines are bulged around the flight axes, hence giving the EM character.

The sphere "cars", F-spheres, line up along a centerline in accordance with the mathematical inversion transformation which must occur to transform spacetime curvature from inside the atom to a monotonically increasing time outside the atom, hence the identical multiple car train, no caboose. Since the spacetime curvature spheres line up inside the atom in a geodesic, light travels in a straight line outside the atom as geodesics or modeshapes in elliptic geometry space, must be inverted to outside the atom as a straight train of transitioned harmonic modeshape of an "excited" electron released into the other spacetime traveling at time speed in hyperbolic or parabolic geometry space.

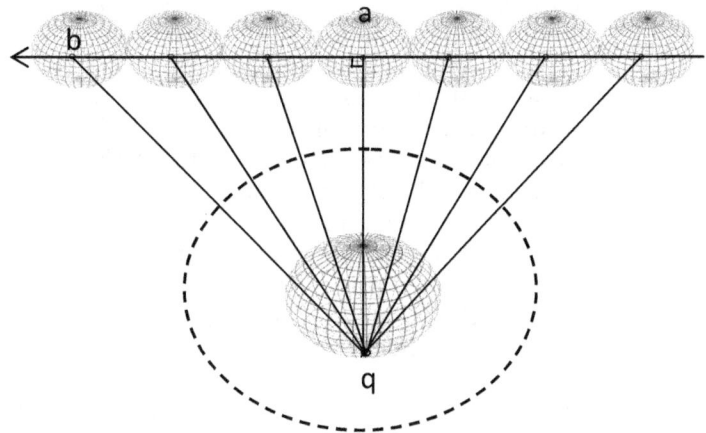

We get there by starting with Mobius transformations and inversion.

$$f(z) = \frac{az + b}{cz + d}$$

Where z is a complex variable, and a, b, c, and d are complex constants. In GR a single 4-vector (t,x,y,z) of spacetime is used in a Lorentz transformation for the complex mapping of what turns out to be a Mobius transformation. It follows also that every Mobius transformation of the complex variable yields a unique Lorentz[41] transformation of spacetime.

> "Thus each Lorentz transformation of spacetime induces a definite mapping of the complex plane. What kinds of complex mappings do we obtain in this way? The miraculous answer turns out to be this: The

---

[41] *Visual Complex Analysis* by Tristan Needham pgs 122-128.

complex mappings that correspond to the Lorentz transformation are the Mobius transformations! Conversely, every Mobius transformation of the C yields a unique Lorentz transformation of spacetime. Even among professional physicists, the "miracle" is not as well known as it should be. The connection is deep and powerful. Just for starters, it means that any result we establish concerning Mobius transformations will immediately yield a corresponding result in Einstein's Theory of Relativity. "

Nature performs a Lorentz transformation point by point from one spacetime position inside an atom to another position outside of the atom. Hence the conversion of one spacetime to another, as we proffer that EM represents, should conform to the physical transformation and the mathematics that models transformation faithfully. To understand the physical transformation between the inside of an atom's spacetime and the outside of an atom's spacetime, we begin with circles, because circle inversion on the radial dimension r, is simplest and generalizable to sphere inversion for three-dimensional space.

We could ponder the connection between mathematics and physical reality. One might wonder if the mathematics is sufficiently advanced to describe this physical observed reality, as spacetime transformation across geometries of spacetime. And not only spacetime but the curvature in the spacetime entity. Flat spacetime, a two-dimensional plane in 4D, is not interesting from a forces and fields standpoint. But curved spacetimes provide ready forces, fields, and "mass", resistance in time and space. Here purely geometrical mathematical entities transform to actual physical entities with forces and fields under curvature alone. The mathematical transformations from circles inside of circles transformed into rays or straight lines or larger circles outside the circle are useful as model validations for a physical model where the two-dimensional is scaled up to four dimensions. These are mathematical underpinnings of our interpretations of physical reality which model what would happen in the physical reality of transformation from one type of spacetime to another type. Moreover, they are consistent with GR and allow us to bring the heavy mathematical machinery inside the atom, to understand what it is we see and measure. But as Tristan writes, "it means that any result we establish concerning Mobius transformations will immediately yield a corresponding result in Einstein's Theory of Relativity. "

    More interestingly to us, spherical harmonic modeshapes of spacetime curvature inside of an atom of a vibrating sphere of one-sided curved surface space are mapped to time-continuous F-spheres outside of an atom, to another spacetime, just as circles are in the explanation above. In this fashion circles, angles and symmetry are preserved mathematically and physically through a transformation between spacetime curvatures. Thus it remains to be proven that this character property of spacetime dimensional mapping holds in physical reality, mapping space curvature inside the atom to outside the atom, and taking into account the 4th dimension of time from frequency to monotonic.

Addressing the Mott problem, how does the creation of linear tracks of photons arise from spherical waves?

In quantum mechanics, the Mott problem is a paradox that illustrates some of the difficulties of physical understanding from a wave collapse and measurement in quantum mechanics. The problem was first formulated illustrating the inconsistency in Quantum Mechanics of the "collapse", physical reality from theory, of a spherically symmetric wave function into the linear tracks seen in a cloud chamber.

Virtually all high-energy physics experiments, such as those conducted at particle colliders, involve inherently spherical wave functions. Yet, physical manifestations of a particle collision detected, are invariably in the form of linear tracks and it matters not what type of particle collider is used. Of course, QMers have some explanations, none of which are persuasive and only point to the fact that QM is not a complete model, and dancing explanations defying logic is the path taken to preserve and prop up a bad theory from the beginning.

A much simpler and elegant explanation is provided by the mathematics of inversion transitions between spacetime describe here and above. Where a straight line in one spacetime must map to a straight line in another spacetime, and the transition will change the spacetime curvature in the process. The change of spacetime curvature is always measured and or observed in terms of energy, but the physical manifestation from a strictly physical model explains the linear tracks typically observed from particle decays as a result of the geometrical preservation of geodesics, aka straight lines.

Our model applies a recursive approach of geometry on spacetime, to explain an unexplainable physical manifestation of EM, atoms, subatomic "particles", light-matter interaction, and EM's, straight-line trajectory. The physical ramifications of the model where photons transform from within atoms or electrons and vice versa, to propagate in Mobius or dimension Inversion transformations conforming straight line, not along infinitely many probabilistic paths that converge to a straight path[42], the current explanation from QED.

Moreover, this model provides conservation of lines principle known as "preservation of circles", a predecessor or mathematical foundation for our conservation of spacetime curvature.

A most interesting corollary of these Mobius and Inversion transformations of circles within the unit circle is found in the "problem of touching circles"[43]. That is where complex inversion maps sequential touching circles between a larger circle A and an inside circle B touching tangent to one side of the circle A, inside circles of progressively increasing size that are touching and are at right angles,

---

[42] *QED – The Strange Theory of Light and Matter*, Richard A. Feynman, pg 53

[43] *Visual Complex Analysis*, Tristan Needham, pg136-137. "In [14] we have drawn the unique circle K centered at q that cuts C3 at right angles. Thus inversion in K will map C3 to itself, and it will map A and B to parallel vertical lines; see [15] "

harmonic modeshapes in series, to another circle K, the touching circles, each of increasing harmonic. The sequentially smaller but touching circles, mode shapes of the harmonics, between A and B are transformed to a straight centerline for a circle stack outside of circle A, defining the trajectory of an aligned circle "train" where the train circles are transformed from progressively increasing to of equal size.

As we know Klein bottles have a one-sided surface and two Mobius edges. Hence a series of Klein bottles connected end to end forming one long one-sided surface for stretching out in the macro time dimension is our first physical step in transforming from one spacetime to EM of another. Below is an image of that sequence of Klein bottle objects connected, in what would physically be a single-sided curvature twisted and rotating surface having both positive and negative alternating curvature.

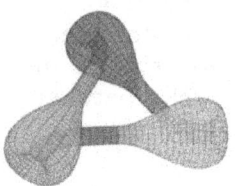

The photonics community is just recently coming around to understanding the topological opportunities in the study of photons and photonic interactions. With the Chronature model of photon structure, we start by connecting up morphed FSphere Klein bottles whereby we can provide a physical EM model comprised of a one-sided topological surface sphere like a train, which represents the photon transformed from a region of spacetime inside of the atomic spacetime with frequency-time, to a ray-like, physically a sphere train outside the atom or electron in straight time. This gives us the wave and the particle character, that duality nature of photons. Hence the structure of a photon is not the classical two perpendicular 90 degree shifted and phased waves, but a one-sided surface sphere train of EGSFT transformed to HGSMT spacetime curvature, where the Klein bottlenecks are moved to the "sphere-like poles" and reduced in size of the bottle "drinking" opening and "handle" to accommodate a one-sided surface sphere-like structure having reciprocating negative and positive curvatures.

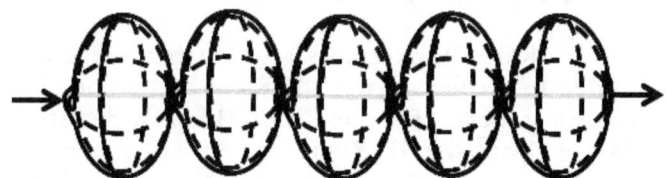

We know that EM has a property called circularity. We can create right-handed circular light or left-handed circular photons. We can also build filters to get only one or the other circularity. How do we explain this from our physical model of a photon? A topological property of a Klein bottle that corresponds to its one-sided nature surface is a property called "non-orientability". What this physically means is that on a Klein bottle topology one cannot covert the surface left-handedness from right-handedness or clockwise rotation from counterclockwise rotation without "sliding" up or down the one-sided surface from one side of the time dimension throat section to the other. As it turns out this feature is remarkably and naturally depicted in our morphed Klein bottle F-sphere train model

What is the physical nature of EM?

We are taught that the "B-Field" and the "E-Field" are physically perpendicular and adhere to left-hand and right-hand rule circularity respectively or why they seem somehow bound to each other. But we proffer that the "B-Field" and the "E-Field" are manifestations of spacetime curvatures dimensions perpendicular or orthogonal to each other and any point along with their geometry.

"Right-handedness and left-handedness were alluded to previously wrt the Klein bottle one-sided surface property. For the E and B fields, picture the EM wave with the 4D Klein bottle F-sphere car train model. Each inverted F-sphere in the sphere train has a "south" or negative curvature outside attracting pole and a "north", positive curvature on the same one-and-same side surface repelling pole, not unlike shown in the bar magnet Minkowski diagram previously. These two opposing curvatures on the same-sided Klein bottle surface of a morphed spherical configuration provide the "E-Field", attractive force or negative curvature, and the "B-Field", positive curvature repulsive field. This model reconciles the classical photon or EM wave Maxwellian B-Field and the E-Field forces with a manifestly physical spacetime model and solely through spacetime curvature of a magnetic south pole attractive curvature and magnetic north pole repulsive curvature on the same sided surface but 360° cycles. But how are the E-Field and B-Field physically perpendicular in our Chronature model? The reader will note that as the "handle" of the Klein bottle enters the "inside bottle" volume, still the same surface side, there is a dimensional shift, a Mobius twist, rotating the surface 180°. The actual complete cycle is then 720° since it's a Mobius twist from outside to inside, but traditionally and classically collapsed to 360°. Hence the EM train F-spheres cycle and rotate the "E and B fields", time transferring from North to South cyclically, in propagation speed C, the time cycle component from its originating matter spacetime.

As a reminder, the F-sphere has a spatial rotation about its axis, the entire photon F-sphere train, having spin and rotation symmetry about the translational axis, giving the properties of left or right circularity. Hence our physical model of EM F-sphere train manifests the dual physical properties of wave and particle with cyclical "E-Fields and B-Fields" perpendicular to the time travel dimension and each other spatial dimension, as predicted in the current EM Maxwell model.

The photon sphere car train only shows half of the one side of the photon curvature in the 3D depiction. The photon curvature showing the time dimension to the timing component opens and manifests more like the figure below. What is not shown well is that the surfaces are more acutely curved to the inside and axis in opposing cycles, and these provide the repeating positive and negative surface curvatures currently only known as the E-Field and the B-Field.

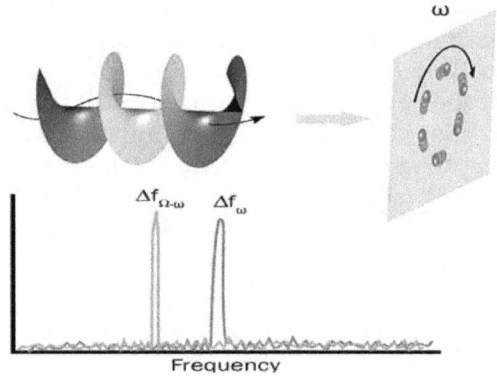

This begins to look like an alternating edge warped classic Rieman Surface, with an elliptic function meromorphic on C, complex plane, for which there exist two non-zero complex number ω_1 and ω_2 with ω_1/ω_2. not elements of C, such that

$$f(z) = f(z + \omega_1) \text{ and } f(z) = f(z + \omega_2) \text{ for all } z \in \mathbb{C}.$$

This may be of interest when we attempt to transform frequency-time elliptic space into macro spacetime, as would be manifested by a photon.

How does light's propulsion work to accelerate to and stay at light speed?

A typical answer given today is that **light** is always moving at c because **everything** is always moving -- or would be, if it had its way around matter, ie. the default speed of spacetime and hence the universe is light speed c. If there's nothing to slow you down, there you are, moving at light speed c right along with the rest of the universe.

So...that begs the question, why are **you not** moving at the speed of light if spacetime's default speed is lightspeed and you live in spacetime? The answer is we are all moving at the speed of light, but through the dimension of time or at the speed of time c. When we attempt to travel faster through spatial dimension, we must give up some of our speed in the time dimension. That's why time ticks slower as we accelerate through space, as space and time are monolithic and the constant in spacetime or space and time are combined.

In addition and as we travel through space we are subject to the warp of spacetime made by matter embedded in it and our interaction with matter slowing in space and time is called inertia. All particles interact with something called the Higgs field, it has mass which has a lot of cool properties – but it requires "energy" or spacetime curvature, to get you moving to light speed in space. The only things that can move at that light speed through space are things that are already going at that speed because they have not to mass -- and therefore don't have any interaction with the Higgs field to give them inertia.

But why do massless things move at the speed of light in the first place?

Physicists tell us that a photon is a massless particle. Photons don't interact in a meaningful way with the Higgs field, in a vacuum, at least, to slow them down. So light, EM, just runs at its speed, which is the speed that everything in the universe would run if it didn't have matter of any kind getting in the way and it was not using any of its time allotted dimension while in spacetime. Another explanation is that it's not that massless things move at the speed of light; it's that we call things that move at the speed of light 'massless'. That's how they define what 'massless' means. This all sounds hooky to me too.

So what really propels light or why does light fly and so fast?

Spacetime curvature provides EM wave propulsion. None of the current definitions above really answer what and how is light propelled, either physically or theoretically," massless" or "universal speed default" notwithstanding. The short answer for how is that photons are born from atoms. They started as standing wave vibrations on an atom boundary, vibrating on a spacetime with a rotating time dimension and we call these standing waves "electrons." Under certain conditions, these standing waves in the rotating time dimension convert to traveling waves in the monotonically increasing time dimension. The spacetime inside an atom, picture a ball of yarn representing the rotating time dimension, unravels the circular time $1/c$, into a linear time c in the outside spacetime, picture holding a string of yarn at one end and throwing the ball of yarn, and is manifested as photons traveling at the speed of time, c. But it's all the same thing, spacetime curvature, manifested in accordance with the spacetime geometry where the curvature lives.

How does a photon propel itself and why does light fly at light speed c?

First let us dispense with the mathematical Maxwell model of a traveling EM wave, where the E-Field and the B-Field alternate sinusoidally in time. In physical reality, the E-Field and B-Field are orthogonal dimensions with spacetime curvature train shaped in spherical cars rotating synchronously in space on a spacetime wave. The sphere surfaces are the spatial dimensions and are measured as changing E-Field dimension orthogonal to another spatial dimension B-Field, the dimensional spacetime curvatures orthogonal rotating and moving at the speed of time through the 3rd spatial dimension. This process of moving curved spacetime propels the wave train. The Maxwell model is strictly symbolic, with no relevance to the actual physical propulsion of a photon. The governing force acting on a single charge is known as the *Lorentz force* and is given by

$$\mathbf{F} = q\mathbf{E} + q\mathbf{V} \times \mathbf{B}$$

where a force **F**, or acceleration, is need manifests as a on a charge q, an asymmetric particle or vibration with velocity is **v**, the instantaneous velocity of the electron particle. However there is no real "particle" here as the photon is massless, and no charge as the photon is chargeless. Hence there is no force from an E-field or the B-Field from the Maxwell model.

Now the Chronature model, starting with the attractive and repulsive spacetime curvature structure creating a magnet, we multiply that structure of an F-sphere operating in an F-sphere train as shown from the mathematical Inversion operation in the atom spacetime. In a stationary magnetic field diagram, we can see that the magnetic forces of the south-pole and north pole arise from the spacetime warp as the F-sphere curvature changes from the inside to the outside of the one-sided F-sphere surface going from negative to positive curvature respectively. This curvature change occurs relative to the centroid of each F-sphere. The one-sided surface then cycles through an attractive force towards its centroid and then a reverse repulsive force against the centroid of the curvature surface, pulling forward and pushing back on the F-sphere centroid surface. By analogy in the macro world, this is like Bernoulli fluid flow over an airfoil surface exerting a lift and reverse drag forces due to pressure differential on the surface. With photon transport, that fluid flow is moving time flow and the "airfoil" is the F-sphere one-sided surface.

Please notice even though we have used an aerodynamics analogy to illustrate forces on a curved spacetime surface, nothing more than the geometry of spacetime with a moving time dimension causes these the very physical acceleration fields on an F-sphere, a physical object created from spacetime curvature boundary, propelling it forward in macro spacetime at the speed of time c.

Again, as the curvature of the F-sphere attracts its boundary to the centroid of the F-sphere surface, a force is exerted that pulls on that surface towards the centroid point. Upon reaching the centroid in time, the curvature switches direction and flips in time to repel the centroid of the F-train surface. This pulls forward - push back type of spacetime dynamic propulsion gives it the wave and circularity character of a photon. For a physical depiction see the spacetime curvature figure below. The packet character comes at collisions with matter. As mentioned above, the smaller radius F-spheres, the smaller the wavelength, the more F-spheres or higher the frequency, and hence the larger the spacetime curvature traveling. Hence the smaller the F-sphere car, the larger the frequency and the higher the "energy" upon the impact of the train of spacetime curvature upon collision with atomic or matter spacetime boundaries, which will absorb and or deflect the wave train.

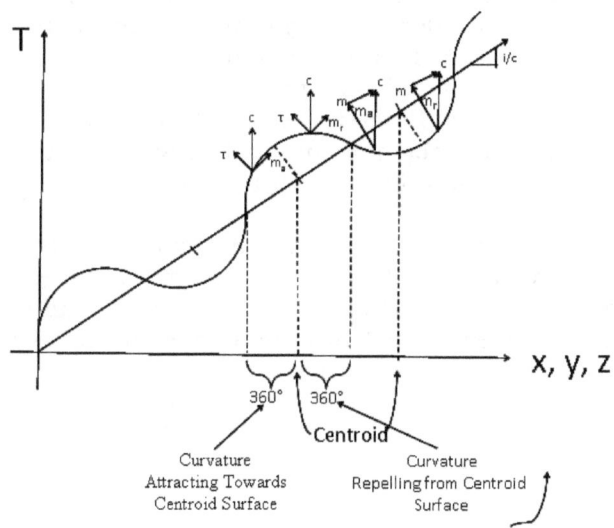

Please note that the photon travel locus extends beyond the world line light cone. This is known as the wave phase velocity, which at times exceeds the speed constant c, but whose group velocity, the average wave train, does not, where the multiple of phase and group velocities equal the maximum speed constant squared, c^2.

What are "quanta" in photon packets?

Plank did the heavy lifting in measuring the relationship between energy E and EM wave frequency f. Planck calculated that the number of quanta carried by an EM wave was proportional to the frequency of that wave, ie. the proportionality between the "energy" and frequency. Einstein's Photoelectric Effect showed that the greater the frequency of the light the higher energy of electrons, volts, being emitted from the surface absorbing the light frequency. Einstein reasoned that light consisted of packets of energy. But what are these "packets" or quanta represent physically?

A visual inspection of two equal-length photons shows that for a given length C of the photon train, the smaller the sphere radii and the more spheres in the train the higher the spacetime curvature.

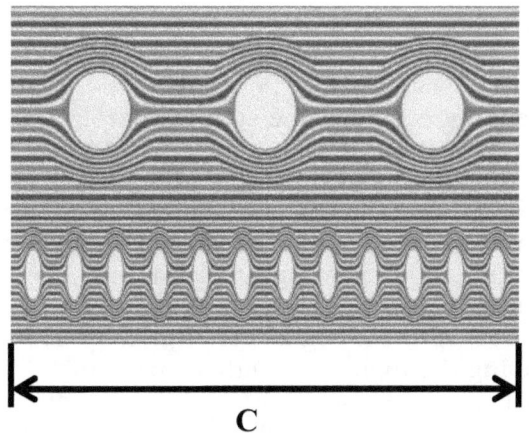

C

From Plank we have his constant of proportionality **h** between photon "energy", **E**, and it's Frequency **f**:

E = h*f   = Planks Constant * Frequency
E = h*f / λ   = Planks * Frequency / Wavelength

How can we get to the same place as Planck, only with Chronature spacetime curvature application? By inspection, the total curvature of the photon train of length c could be calculated as equal to the sum of curvatures of each sphere on the EM train, or C times the number of spheres, n, for the same time interval. Thus

$$E = C_n * c/\lambda$$

where **C_n** is a constant comprising the individual sphere car curvature for frequency n, c is the length light would travel during an emission, and **λ** is the length of one sphere car. Writing the constants as **h**, we get the same equation as Plank;

$$E = h*f$$

where energy is simply the sum of spacetime curvature involved in spacetime curvature collision. So, this diagram also shows why the photon "packet" of "energy", is simply the total spacetime curvature of the F-sphere train interaction. Although spacetime fluctuations or alternating acceleration fields, photons also come in packets, total spacetime curvatures, with the higher "energy" F-sphere trains also having the higher spacetime curvature fluctuating acceleration field "cars" and therefore impact impinged matter with more "energy". So again, the "energy" of a photon is related to the number of F-sphere cars, fluctuating acceleration fields or changing curvature, in the photon train in its total train spacetime curvature, comprising a packet or "quantum of energy" as they are transformed in discrete harmonics at the atomic boundary.

Modernly, scientists use Wavenumber, which is roughly the number of sphere cars on an EM train per unit length. Sometimes that's a better way for measurement and understanding.

From this, it would appear that the actual space curvature of each sphere car is the same regardless of the sphere frequency. This would stand to reason since each sphere car's curvature has equal amounts of negative and positive curvature which would zero out each other in some sense. But the positive and negative curvature amounts to magnetic force, the two-pole spherical structure of a magnet described above, repelling and attracting curvature on an impinged object in a way to transfer acceleration-deceleration to the object at the objects next available, unexcited, harmonic frequencies. Therefore the photon train with more cars with higher curvature would hit with more impact than trains with lesser F-sphere cars with less curvature, remembering that higher radii are inversely proportional to spacetime curvature magnitude. The quanta of n cars in the train would then be larger with the larger number of cars, higher frequency, more "energy". This physical model predicts the Planks constant of proportionality with the frequency of cars and energy impact transferred of EM "wave".

What happens when photon trains "collide"?

Current theory teaches that photons can be rudimentarily conceptualized to be "split" in a process known as spontaneous parametric down-conversion, SPDC, which is essentially the inverse of second harmonic generation; the process by which for example green laser pointers produce their visible light by summing two infrared photons. Of course, photons are fundamental particles, so the only thing we get out when they're "split" are more photons.

Photons are currently labeled as "mediators of force", so splitting them results in smaller energy versions of themselves. Moreover, two "low energy" photons give you one "high energy" photon and the reverse is also true. Some materials have the necessary properties to manipulate photons in this way. Photonic wavelength doublers are an example used for green lasers. During these conversions, the total amount of energy stays the same.

What does this all mean in the Chronature photon physical model of the individual sphere curvature light-speed moving sphere train? In experiments, an original photon and a suitable crystal lattice, are "phase-matched" in the frequency domain, so that they have "correlated polarizations". The measured phenomena would suggest that the links between the photon individual spheres can be severed and space curvature recombined with other phase-matched photon sphere trains but only with "correlated polarization", whatever that means physically. Since photons all travel at the same speed the down-conversion mathematical modeling process should be less time-dependent and more spatial harmonic dependent. Thus the "correlated" photons would add or if their spacetime curvatures were "vertically-polarized" and "horizontally polarized" to coincide with the direction of space warp and thus combining. All this means in our physical model is that for photons to add or subtract, their spatial dimensions must line up. Taking a physical example from water surface waves interacting at the head-on or perpendicular directions; head on the waves are additive, perpendicular the waves are independent of each other's travel direction and interact differently, and not in an additive fashion. So "vertical" or "horizontal" polarization translates to spacetime dimensional alignment between photons.

Diffraction and Diffraction Patterns

No chapter on light and matter interactions would be complete without addressing the single and double-slit diffraction experiments. These are described and treated well in most physics texts in a classical way. According to QMers, these experiments prove that light, and all EM, have wavelike, diffraction patterns, as well as particle-like probabilistic characters that are quantized. The issue overlooked is the explanation of the physical nature of this light-matter interaction and the actual structure of light. The wave versus particle property of EM and matter has always been particularly impossible to explain with the current theories. However, if light were a spherical car train of alternate spacetime, its interaction with matter would indeed be in the form of a diffraction pattern that we find.

A relevant observation of light scatter is the water droplet experiment[44], where the droplet surface acts to produce different spherical wavefronts.

---

[44] Atmospheric Optics, "http://www.atoptics.co.uk/droplets/corform.htm"

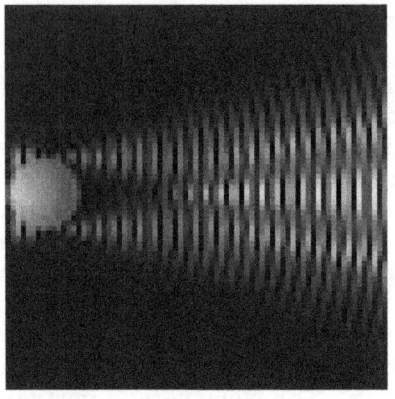

The above diagram illustrates how two points on a spherical surface scatter light and act as sources of outgoing spherical waves. The scattered waves overlap and interfere. Where wave crests of the same sign coincide the light intensity is increased, in the same way, that a spherical wave train of spheroid cars would disperse. These go a long way to describing the two slot experiment results, where the fluctuating sphere train in spacetime have opposite amplitudes they destructively interfere to give low intensity. Scattered light from the whole sphere's surface plus smaller contributions from reflected and transmitted waves combine to form a diffraction pattern. This interaction may explain the two slot and single slot experiments proving that the "duality" of light and matter manifests as particles or waves depending on the interaction and transformation from two different spacetimes, the macro and the atomic. In addition, the interaction may be affected by the slot edge scattering character.

Under Huygens' Principle, every point on a wavefront can act as a new source of waves. Therefore electrons impinging on matter at the slot walls can also cause the diffraction patterns by scattering off the wall atoms. Sending one photon or electron at a time makes little difference, the pattern on the wall will be the same.

The diffraction pattern from a droplet is almost the same as that from an opaque disc or even a soap bubble. In turn, the diffraction pattern from a disc is the same as that from a circular aperture of the same diameter. This aperture can be an atom or molecule as described in Chronature, with the physically vibrating boundary of spacetime curvature.

Also, a photon sphere train as described in the Chronature model above can well produce the diffraction pattern without regard to whether it was a wave or particle, but simply a sphere train of connected rotating spheres of oscillating spherical space curvature traveling at relativistic speeds, interacting with atom 3-sphere spacetime, vibrating spheres of a different spacetime but with spacetime curvature which directs the photon train into a diffraction pattern. Hence it matters not what the mass of the traveling sphere train, or even if the sphere train were treated as a particle, the effects would be the same and observed diffraction accordingly.

Compton Scattering and others

One of the most important discoveries in physics regarding the interaction of matter and EM is the Compton scattering characteristics of EM off of matter, generating free electrons and a reduced frequency of reflected EM. What is happing from a standpoint of Chrono Spatial Dynamics (CSD) and how can we take advantage of a purely geometric way of understanding this important physical phenomenon? Moreover, Compton is the "mid-energy" phenomenon, with the "low-energy" being the Photoelectric effect and the "high-energy" phenomena called Pair production. If we can show why these kinds of interactions divide up into these three phenomena. The figures below depict the Compton scattering as physics currently presents it:

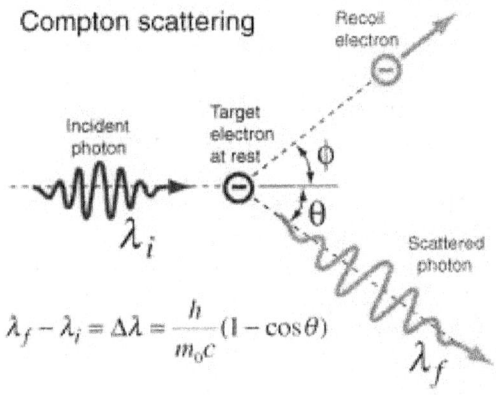

Photoelectric Effect:

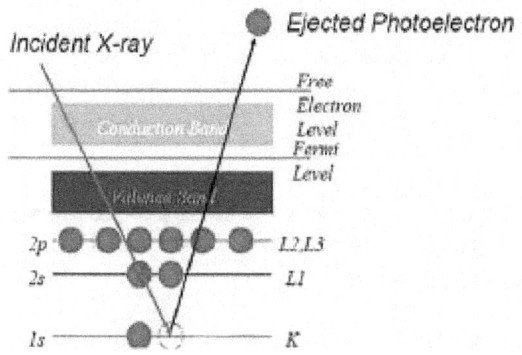

Pair Production: Matter – Antimatter production/consumption shown above as Feynman Diagram.

The Compton Scattering occurs from x-ray frequency(30×10^{15} Hz to 30×10^{18} Hz), wavelength (0.01 to 10 nanometers), and amplitude EM. From a dimensional standpoint, these parameters are near the physical dimensions of the target atoms, 10^{-8} cm.

From the classical atomic mechanics model, the atomic angular momentum must be conserved to a total from the sum of the electron orbital angular momentum and the spin angular momentum. In the Chronature model, it is possible that the incoming photon transfers spacetime curvature into an atom to the extent possible but at only structures natural frequencies or harmonics at non-excited frequencies and without duplication of existing harmonic frequencies, transfer its spacetime curvature into an atomic shell without a nucleus, like an electron. The remaining spacetime curvature if insufficient to contribute to another available natural frequency atomic spherical structure, then the remainder is shed as a lesser frequency photon and leaves the interaction at the Compton angle.

How are light-matter interaction phenomena explained in the Chronature model and what is physically happening?

In the case of Compton and Photoelectric Effect the incoming spacetime, "electrons," must have sufficient spacetime curvature to transform to the minimum discrete harmonic spacetime available, within vibrating sphere harmonics and molecular resonances. In terms of Chrono Spatial Dynamics, this is analogized to an incoming traveling spacetime wave colliding with an array or network of standing wave spacetimes.

The harmonics of the standing wave network, atomic lattice, as we know and measure them are electron levels, but for us, these are discrete harmonics of the atom spacetimes containing surface membrane boundary. The incoming EM can be modeled as alternate spacetime forcing function with frequency content acting on the atom harmonic frequency levels, the natural frequencies of the impinged structures comprised of spacetime curvature. Only harmonic frequencies of the atomic boundary surface or membrane will be allowed to transition from one spacetime to the other and at the resonant frequencies. That means if the spacetime curvature frequency of the incoming EM is lower than the fundamental mode frequency of the atom boundary, the EM will pass without any transition, similar to a wave passing a structure without resonances that will have no interaction. A frequency of incoming EM spacetime curvature perturbation will only excite atoms natural frequencies which are lower because the incoming frequency content will deposit at the discrete frequencies of the atom boundary spacetime curvature with the remaining frequency content moving on in the spacetime of origin.

Another way to explain this is from wave fundamentals. When two undulations from different origins but changing spacetime curvatures meet, their amplitudes sum in combination. That is to say that at the atom boundary, the incoming EM wave amplitude, additive curvature, is synchronized with the amplitudes of the atom boundary vibration standing wave harmonic amplitude curvature at their crests effectively adding them to the atom resonant harmonics. Remembering that the atom surface boundary is a sum of the periodic spherical harmonics of the atom sphere vibrations. Thus incoming spacetime curvature harmonics combine with resonant atom harmonics, "electrons." The combined spacetime curvature amplitudes at the resonant, "electron," harmonics then must be transferred back out to the incoming spacetime, due to the Pauli principle as explained above.

In the Chronature model, Compton Effect is an interaction of spacetime solitons with a lattice of atoms, where the impinging soliton wave emerge with the original shape, speed, and size with the exceptions of the spacetime curvature conversions between two complex waves represented in two different spacetimes, partial conversion of soliton to the atomic spacetime and residual curvature less curved, "lower energy". Oddly enough, soliton interaction solution techniques[45] were borrowed from techniques used to solve the Schrödinger equation. Soliton researchers found through computer simulations of soliton waves that a "train of eight subsidiary waves, which separated from the original wave and then coalesced with it again.." In other research on solitons "if a pulse of energy was applied to one end of the network, it flowed through all the nodes and took on a complicated and seemingly random distribution; after a time, however, the pulse seemed to reassemble itself and return to its starting point." In our analogy, the "network" is the atomic lattice and the "energy" is in the form the incoming EM soliton photon.

In brief overall, this means that if f_{in} is > f_{spfd} then $f_{in-spfd}$ can transfer that discrete amount of curvature to the available "electron" harmonic, and the remaining frequency will not be translatable and must pass on in the spacetime of origin, frequency prime. Since the atom membrane already has an existing harmonic and it can only have one, it must then shed that in the form of a free electron, a spherical shell of spacetime with the one harmonic. Frequency jumping from one electron shell level to another "lower energy" or lower frequency and emitting a photon means that the atom membrane can transition from higher frequency content to lower frequency content because the curvature of higher frequencies is larger than the curvatures of lower

---

[45] Solitary Waves" from Scientific American, Volume 80, pg 358

frequencies and a remainder curvature will configure in the spacetime of origin, staying in that spacetime with the remainder curvature.

Another way to see this is to remember that sphere shell harmonics go by $1/n^2$ where n is the harmonic number. The higher the harmonic the tighter its corresponding modeshape and hence, the larger its spacetime curvature. Hence the larger curvature can form smaller curvatures inside the atomic membrane and emit the remaining spacetime curvature in the form of EM spacetime at another phase or angle. Thus spacetime is conserved.

Where this logically leads us is to better use of tools. We currently have femto pulse lasers, the capability to shape pulses and sequence various wavelength EM over a range of wavelengths never before possible. One titanium-sapphire laser can operate within 10-20 nm bandwidth with 100 femtoseconds (fs) pulses. Some broadband lasers operate within the 100 - 200 nm bandwidth with 10 fs pulses. Imaging, beam propagation, repetition rate modulation, spectral phase, and amplitude shaping, and adaptive spectral control are providing scientists much better more precision tools but the current models are weak or non-existent and cannot take advantage of computational geometry methods and capabilities without undue experimentation. Numerical modeling using the Chronature structure of photons and matter interaction will help, by providing clean elegant models which are computationally relatively simple wrt to current QM numerical modeling technology.

What is matter, that it would interact with light?

Under the Standard Model, the Higgs Boson is posited to have been the agent that gives mass and energy to matter after the creation of the universe - AKA the "God particle." But what is mass and energy of EM in spacetime curvature dynamics? Above we posited that mass is related to the dynamic viscosity of spacetime curvature based on the model equation placement of the mass parameter m. The reality is that we observe the universe of mass and matter in a sparse distribution, whereas there is much "space" between the globs of matter in the form of astronomical bodies. We also previously defined "electron energy levels" to spacetime curvature differentials to show analogies to current Rydberg atom models and also in Chronature as:

$$dE = K_n(1/r_1^2 - 1/r_2^2)$$

give or take a constant, where classically the "r"s are analogous to the principle quantum numbers which represent the orbital, but in Chronature model represent the radius of curvature for the modeshape harmonic, which are the actual physical radii of the "electron orbital" or modeshape. But in both models, the changes in "energy", transformed spacetime curvature from one spacetime to another, are proportional to the inverse square of the orbital radius of curvature or inverse square of the principle quantum number, in the Rydberg model. So the answer to the question of what is matter that would interact with light is that the interaction between light and matter is an exchange of spacetime curvature, which fits by analogy with measured data of such interactions in the Rydberg series.

Wherein the Rydberg model the "energy" of light emission or absorption results from an electron jump between the second energy orbital closest to the nucleus and those orbital more distant, in physical reality Chronature model the photon is a traveling oscillating spacetime curvature that adds or subtracts spacetime curvature from a vibrating 3-sphere atom boundary in harmonic increments. The harmonic increments occur in the set of natural frequencies of the atom spherical harmonics whose modeshape radii can be calculated from vibrating sphere harmonic solutions to obtain the frequencies and energies of interaction.

Another way to understand this is, light is a spinning sphere train of spacetime that's also vibrating transversely to the direction of travel. When, say the horizontal metric is contracting, "E-Field", the vertical metric is expanding, "B-Field" simultaneously while the transverse plane is rotating or spinning as the wave is moving in the transverse direction. This is much like a gravity wave but with orders of magnitude greater spacetime curvature.

What is the "mass" of a quantum of light?

Let's start with some basics like the classical definition of EM energy, $E = h\nu$, and $c = \lambda\nu$, and then add the spacetime curvature of the above model of EM, a train of spheres. A typical EM wave shows the EM wavelength is in the same order as the wave amplitude which we will use as the radius of curvature of the EM spacetime. Combining the "energy" E and using the radius of spacetime curvature R_1 as ½ λ we obtain v=2c/R or R=2v/2c. We define energy as spacetime curvature,

$E = \Sigma (K/ R_1^2) dt$, and equate that to the classic EM energy at a surface point solely due to spacetime curvature. The dot product with a control surface dA over the entire surface gives us $4/3 \pi R^2$

$E = \Sigma (hv) dt = K/R_1^2 = K4v^2/c^2$

v

$E = h^2c^2/4K = mc^2$

rest mass $m = h^2/4K$

So it would appear from a strictly Chronature point of view, that EM's spacetime curvature does not have much to do with the EM's rest mass is a constant equal to $h^2/4K$.

Where is there room for more discoveries into the interaction of light and matter?

Over the years, many Nobel prizes and accolades have been awarded to scientists for discoveries in the EM-matter interaction field. Such phenomena as the Zeeman effect, Stark effect, Compton scattering, Photoelectric effect, Inverse Compton effect, Lamb shift, and others have been promulgated and advanced our understanding of our physical universe. Briefly:

The **Zeeman effect** is the effect of splitting a spectral line into several components in the presence of a static magnetic field. It is analogous to the Stark effect, the splitting of a spectral line into several components in the presence of an electric field. The Zeeman effect is very important in applications such as nuclear magnetic resonance spectroscopy, electron spin resonance spectroscopy, magnetic resonance imaging (MRI), and Mössbauer spectroscopy. It may also be utilized to improve accuracy in atomic absorption spectroscopy.

The **Stark effect** is the shifting and splitting of spectral lines of atoms and molecules due to the presence of an external electric field. The amount of splitting or shifting is called the Stark splitting or Stark shift. In general, one distinguishes first- and second-order Stark effects. The first-order effect is linear in the applied electric field, while the second-order effect is quadratic in the field.

Compton scattering, discussed above, is the scattering of a photon by a charged particle, usually an electron. It results in a decrease in energy (increase in wavelength) of the photon (which may be an X-ray or gamma-ray photon), called the **Compton effect**. Part of the energy of the photon is transferred to the recoiling electron. **Inverse Compton scattering** occurs, in which a charged particle transfers part of its energy to a photon. Compton scattering is an example of inelastic scattering of light by a free-charged particle, where the wavelength of the scattered light is different from that of the incident radiation. In

Compton's original experiment, the energy of the X-ray photon (≈ 17 keV) was very much larger than the binding energy of the atomic electron, so the electrons could be treated as being free. The amount by which the light's wavelength changes is called the **Compton shift**. Although nuclear Compton scattering exists, Compton scattering usually refers to the interaction involving only the electrons of an atom.

The **Lamb shift** is a difference in energy between two energy levels $2S_{1/2}$ and $2P_{1/2}$ of the hydrogen atom which was not predicted by the Dirac equation, according to which these states should have the same energy. Interaction between "vacuum energy fluctuations", another theory, and the hydrogen electron in these different orbitals is the cause of the Lamb shift, as was shown after its discovery. The Lamb shift has since played a significant role through vacuum energy fluctuations in the theoretical prediction of Hawking radiation from black holes.

Bethe was the first to explain the Lamb shift in the hydrogen spectrum. Bethe was able to derive the Lamb shift by implementing the idea of mass renormalization, which allowed him to calculate the observed energy shift as the difference between the shift of a bound electron and the shift of a free electron. The Lamb shift currently provides a measurement of the fine-structure constant α.

Bound or shift electron, in the Chronature model of spacetime curvature, **Lamb effect** is an example of where atomic harmonics of one class of harmonics or shells, eg. S shell, having specific spacetime curvature at tightest spacetime curvature points on a curvature that is nearer the harmonics of another shell class, eg P shell, having tighter spacetime curvature on harmonic mode shape, Lagrangian from geometry or "energy level", exceptions to the general progression predicted by the Aufbau approximation or Madelung Rule, electron shell progression model. The spacetime fluidic structure interaction here is the spacetime curvature match up aka minimum "degree of degeneracy" eigenstates from QMT, where the harmonic curvatures in Hydrogen S and P shell surfaces at certain mode shape surface points are minimum to the next progressive harmonic in the s, p, d, f harmonic class modeshape. The current shell progression models are n, principal, and l, azimuthally, a quantum number based and atom singularly based, only approximately related to surface curvature values and do not account for all spacetime effects like "shift" and "renormalization", concepts brought in by Bethe.

These are all scientific discoveries and findings with explanations needed for a deficient physical model. The downside played in all of these, is that without a good physical understanding, physicists are forced to rely on probabilistic and statistical theories models from experimental results making them less useful for prediction and applications. And because of their primitiveness or inadequateness, the models are too computationally extensive to probe deeper. So the explanations although supportive of the findings, are inadequate to fully employ.

Photon-matter interaction is entering a new phase as lasers get cheaper and faster. Ultrafast laser pulses are being used to distort the properties of matter and generate electrical currents faster than in any traditional way along with tiny, nanoscale, electrical circuits. The behavior of matter is altered when driven far from equilibrium, by simply varying laser parameters. Most of this work is currently done absent good computer models, so the work progresses slowly.

8 –Chrono Spatial Dynamics and Metric

The most recent great idea of physics theories today is the extra dimensions in spacetime and the theories that support them in the deliverance of a GUT. A half-hearted search goes on for those extra dimensions but only because the TOE model dimensions are too astronomically tiny to ever measure. These take the form of looking for subatomic particles which exist or at certain energies as predicted by the Standard Model. There has been much speculation as to smaller and curved dimensions but so far nobody has been able to physically find even a single extra one, apart from the four dimensions that we know. Their proponents argue that the higher dimensions are very tiny and it will be forever before we can even tap into or measure them, so we must just believe or take on faith that it's beyond us mere mortals to physically understand how Mother Nature magically hide all to extra spatial dimension out there in the universe.

But GR brings us to and leaves us with only 4 dimensions and these through mathematics alone lead to the Einstein field equations (EFE). These may be found written in the form:

$$R_{\mu\nu} - \tfrac{1}{2} R\, g_{\mu\nu} + \Lambda g_{\mu\nu} = \frac{8\pi G}{c^4} T_{\mu\nu}$$

where $R_{\mu\nu}$ is the Ricci curvature tensor, R is the scalar curvature, $g_{\mu\nu}$ is the metric tensor, Λ is the cosmological constant, G is Newton's gravitational constant, c is the speed of light in vacuum, and $T_{\mu\nu}$ is the stress-energy tensor.

In the literature, you will find that the EFE is a tensor equation relating to a set of symmetric 4 × 4 tensors. Each tensor having 10 independent components, but the four Bianchi identities reduce the number of independent equations from 10 to 6, leaving the metric with four gauge fixing degrees of freedom, which correspond to the freedom to choose a coordinate system. What this all means is spacetime has 4 dimensions and any consistent geometry will work.

How does the Einstein Field Equation fit into the real world?

In the way that EM fields exist using charges and current, EFE is used to calculate spacetime geometry resulting from the presence of mass, energy, and momentum. They do this through the Riemannian metric tensor of spacetime for the configuration of masses embedded in a particular spacetime region. The solutions of the EFE are the components of the metric tensor and the inertial trajectories of matter and EM in the resulting spacetime geometry are calculable using the geodesics, whereby the acceleration of the curve has no components in the direction of the curve surface normal so that the motion is determined solely through the bending of the surface in the geometry that it is embedded.

Moreover, with a varied metric from alternate spacetimes, the Riemannian metric tensor also provides the essential map, metric, and structure for the four known forces in the universe, with some claiming the discovered 5th force, comprising the Higgs Field. Moreover, the metrics and alternate spacetimes, the traditional four, create alternate dimensions. The Riemannian metric tensor structures and consolidates the mathematical description that brings it all together in one elegant method providing curvature by geometry dependent distance between points. Physicists and mathematicians have steadily grown the metric tensor from Einstein's GR 4x4 to TOE's 11x11 metric tensor. So this seemed logically reasonable to do after Kaluza introduced Maxwell's equations into the metric tensor and showed unification between Gravity and EM. He turned the mathematical crank and out popped "unification" between GR and EM. Later adding more rows and columns added "unification" of the strong and weak force. It remains to be seen where the Higgs Field will fall in the stress-momentum-energy tensor but rest assured that theory is coming.

From the expansion of the metric tensor additional spatial dimension rows and column none model physical reality, or at least at a level that can be tested or verified. These theories have all failed to prove that six extra curled spatial dimensions can reproduce all of the particles in the standard model. Furthermore, these all have failed to predict a complete set of particles and this is left up to the experimental physicist taxonomists to find and classify. Why physicists hold these theories in such high regard is beyond me, and informed nobody on the map.

And then there is the Primary Conservation Law, that spacetime is conserved. In Chronature, matter is a spacetime curvature entity, so the tensor containing matter should be considered spacetime curvature. This would lead us to conclude that the EFE is

$$R_{\mu\nu} - \tfrac{1}{2} R g_{\mu\nu} + \Lambda g_{\mu\nu} - \frac{8\pi G}{c^4} T_{\mu\nu} = 0$$

Alas, the higher dimensions of subspace, the curled-up dimensions, are not all in the space that we ponder. The space comprising matter, not the traditional 4D spacetime, has been overlooked. Hence the GUT theorists took the wrong turn by discarding physical reality, the easy fix provided by the Metric Tensor. Klein later explained that the higher dimensions of space were curved around the traditional X, Y, and Z spatial dimensions. Although superficially perhaps a physical explanation for more dimensions, the "curled" up much smaller spatial dimensions would not be physically orthogonal to the existing X, Y, and Z spatial dimensions. Also, Klein's explanation failed to address any complete loop in a time dimension, leaving the time dimension as before. So we are back to Chronature, where spacetime is preserved in 4-dimensional units, where more geometrical dimensions make physical sense yet preserver all we see and measure.

How is nature organized with these disparate interacting geometries?

Moreover, beyond a curve in the time dimension, we postulate that the time dimension is curved via CTC loop sense within atoms. Therefore we need to use the solutions for metrics within the mathematical-physics models of CTC and the Gödel Metric.

The Gödel metric provides an exact solution to the Einstein Field Equations representing the matter density of a homogeneous distribution of swirling "dust particles" or non-interacting points. This is also known as the Gödel solution and has many unusual properties. In particular, the existence of <u>CTCs</u> which provide a frequency-time dimension.

The Gödel solution presents the metric tensor in Lorentzian spacetime. As measured by a "non-spinning" observer riding one of the dust grains. "Non-spinning" means that it doesn't feel centrifugal forces, but in this coordinate frame it would be turning on an axis parallel to a spatial axis. The dust grains stay at constant values of the spatial axis of x, y, and z. Their density in this coordinate chart increases with x, but their density in their frames of reference is the same everywhere.

This is beyond our expertise in mathematics for much more explanation of the physical representation here, but it should suffice that the Lorentzian spacetime, a fourth-rank Riemann tensor is a multilinear operator on the four-dimensional space of tangent vectors, at some point in spacetime, but a linear operator on the six-dimensional space of oriented plane segments at that same point.

Accordingly, it has a characteristic polynomial, whose roots are the eigenvalues. In the Gödel spacetime, these eigenvalues are straightforward: triple eigenvalue zero, double eigenvalue $-\omega^2$, and simple eigenvalue ω^2.

Out of differential geometry, the metric tensor function defined on an atomic manifold is used to define the length of an angle between, tangent vectors in that manifold. A metric tensor with a positive length wrt the metric defines a manifold known as a Riemannian manifold. The metric tensor allows us to define and compute the length of curves on the manifold. The curve connecting two points that has the smallest length, "straight line in Euclidean", is called a geodesic, and its length is the distance that a passenger in that manifold needs to traverse to go from one point to the other in the shortest route. The metric tensor itself is the derivative of the distance function. Thus the metric tensor gives the infinitesimal distance on the manifold.

Fields; EM, Strong Force (SF), Weak Force (WF), are dimensionally various but mechanistically unique, ie, spatial "particles" will move in 3 ways in 3 axis spatial dimensions and a temporal dimension; 1) Simple Oscillation or back and forth vibration, 2) Expansion-Contraction about a point, 3) Rotation. These are all manifestations of temperature. The basic motions in 1, 2, and 3 increase with temperature and decrease with falling temperature, becoming zero at absolute zero. Since in Chronature matter is nothing more than a reorganization of a geometric grid or scale in compliance with a given geometry, any combination of the three motions can be used in creating a stable standing wave pattern or a dynamically stable union through geodesic connection paths where momentum is conserved locally. "Energy" is the moving densification-de-densification of space geometry or curvature in a traveling wave moving a maximum propagation within the spatial standing wave field. EM is a pure traveling space wave, ie space whose spatial dimensions cycle. EM waves are combinations of standing-traveling spacetime waves, and the SF is purely a nuclear conversion from positive curvature circulating time-space to negative curvature monotonic time-space. Modernly, the 3 macro dimensions of space and 3*3 = 9 additional time dimensions give us superstring theory, a 10-dimensional metric tensor. TOE adds another dimension to unify super-stings and those of lesser dimensions. We posit a 12-dimensional metric tensor space; 4 in Gravity (x, y, z, t), 1 in EM (Maxwell, E, and B fields are manifestations of the same dimension are a transition between spacetimes), 4 in Atomic Matter (r, θ, Φ, cω) and 4 more in sub-atomic spacetime, with the negative frequency-time (r, θ, Φ, -cω).

As mentioned in the atomic model above, the PCS boundary can receive as well as transform curvature, and this is all done at harmonic resonances, "discrete energies". But summed together in space, these harmonics also provide mode shapes for the full surface curvature. Much like the sum of the sinusoidal component to any actual simple curve courtesy of Fourier Transform, the harmonic modeshapes comprise the surface boundary and hence surface curvature of the atom. The higher the curvature, the higher the attraction to other spacetime surfaces will be. Hence a non-symmetric positive curvature spacetime will exhibit a preference for an attractive surface. This becomes what we call "charge", or "ionic" valence.

In the tensor mechanics, Chronature populates the Metric Tensor by increments of spacetime subtensors, not straight spatial dimensions. The table below shows four candidate spacetimes comprising our Universe and beyond. The reciprocal of the time dimension, frequency-time, and spatial dimensions are non-Euclidean in the vicinity of any curvature. H, F, and E correspond to Hyperbolic, Flat, and Elliptical Geometry. But the spacetime metrics can change at any point, so the table below is a kind of classification of 4D spacetimes with their dimensional parameters.

Spacetime	Time D	S1	S2	S3
Our Universe	t	X, H,F	Y, H,F	Z, H,F
Opposite Front	$-t$	X, H,F	Y, H,F	Z, H,F
Atom	$1/t$ (ω)	R, E	θ, E	Φ, E
Sub-Atomic	$-1/t$ ($-\omega$)	R, E	θ, E	Φ, E

The table above shows the candidate spacetime configurations in a metric tensor. The Einstein Field Equations are not included here, as they may be also be modified later to conform with the non-standard model.

Michio Kaku is the master of explaining complex advanced physics in layman's terms and so I build on his explanation here. The metric tensor is one such topic Kaku covers in elegant style and I include a modified diagram of the metric tensor from his book Hyperspace[46], to show the differences in the composite spacetimes in contrast to the typical String theorists and TOE proponent metric tensor terms.

Macro Space Time	E M - S T	Atom Space Time – Lorentzian	Bo z - Gl uo n - S T	Nucleon Space Time – Lorentzian
EM - STime				
Atom SpaceTime Lorentzian				
Bozon – Gluon ST				
Nucleon SpaceTime Lorentzian				

    This is mere speculation on my part, but knowing only that what we call "particles" are different kinds of spacetime entities. The Macro spacetime represents the current GR gravity 4x4 submanifold which includes the 3 macro spatial and 1 monotonic temporal dimensions. EM – Spacetime is the Kaluza 5th column addition that unifies EM and Gravity, adding the transition partial spacetime of the transition train EM. In Chronature, Atomic SpaceTime is the 4x4 submanifold that contains 3

---

[46] Michio Kaku, *Hyperspace*, pg 146-147

elliptical geometry spatial dimensions and a frequency-time dimension, hence the CTC Lorentzian Metric sub-tensor. Sub-atomic Spacetime, the elementary particles, is also a 4x4 submatrix that comprises another 3 elliptical geometry spatial dimensions and a negative frequency-time, same dimension of frequency-time by opposite sign negative CTC time loops. But as the negative temporal dimension is just the opposite of the positive temporal dimension, we still need the CTC time looping 4x4 Lorentzian Metric sub-tensor for negative frequency-time dimension and 3 non-Euclidean spatial dimensions. As the EM acts as a transitioning agent from Atom to Macro spacetimes and back, the Bozon-Gluon ST performs a similar function between the Nucleon -Atom transitions. Altogether this makes Chronature a 14 dimensional theory, with 3 temporal dimensions and 11 spatial, there exists potentially other recursively smaller spacetime which adds another 4x4 sub-matrix to the metric tensor for neutrinos. However, the spatial dimensions reside and relate to only certain temporal dimensions. Yet all of the String Theories "hidden" dimensions are there, just not as extra curled up sub-space dimensions, instead "curled up" inside of 3-spheres we call atomic, sub-atomic, and below, recursively. This is represented in the matrix above.

In short, the Chronature submanifold spacetime diverse model explains the extra dimensions in the physical evidence that we see and measure without the need for "hidden" or "subspace" dimensions.

What's the matter with Dark Matter?

Cosmologists have speculated on such things as dark matter to explain the expansion rate of the universe and a few other phenomena in need of a theory. I am not a proponent or even adequately versed in the current theories of the Big Bang aftermath or rates of the Universe's expansion, which I recognize to be only as an amalgamation of assumptions and speculations looking for a consistent theory. But let us take it on faith that the measurements are correct in their prediction on Universe expansion rates. If DM exists, how can Dark Matter (DM) be supported by the theory of Chronature, and explain the speculations from purely geometrical spacetime arguments?

What do we know about Dark Matter?

Physicists can tell how large and how fast galaxies move by measuring the speed at which they orbit, and by measuring the deflection of light through and around them. Furthermore, the amount of "mass" from the stars and gas is only about 10-20% of what is necessary matter to account for the measured masses formed/forming and moving. The remaining missing mass, because cosmologists can't see it, smell it or hear it, or interact with it in any normal way, is called dark matter. That and Dark Energy, which make up 95% of the universe, is that unseen and unknown repulsive force that makes the universe expand at an accelerating rate.

The composition of dark matter is unknown, and several underground particle detector experiments are trying to directly detect dark matter particles, but so far nada. Nothing is known about DM, except that our Universe is not accelerating at a speed commensurate with the matter in it, hence there must be matter that we cannot "see". Meanwhile, another question that arises, is it truly extra mass explaining the observations, or is it that our understanding of gravity or other physical law incomplete? Is our calculation of how much mass is there wrong because we need a different law of gravity for such large distances? Perhaps the general relativity calcs are only pretty good, but not the perfect theory of gravity. Perhaps the matter we see is all there is. The fact that this doesn't look like enough matter is because we are using an inappropriate understanding of gravity to calculate galaxy's movements. Perhaps we are using a law of gravity that works well at interplanetary distance scales but doesn't scale at galactic distances.

  But GR has been tested to a high degree of accuracy and seems to be correct and there doesn't seem to be any reason why long distances would suddenly change the way gravity manifests. Hence the explanation "there is more matter than we can see". Perhaps there are anti-gravitational lenses whereby the calculations miss significant amounts of emitted light. Perhaps there are "atoms" that are not 3-Spheres, and do not react with any kind of EM, hence "invisible" to us.

  Observations show galaxies are rotating too fast at their outer edges if there were only luminous mass. Physicists claim cold dark matter (CDM) is needed in the standard cosmological model to be able to predict a universe that matches cosmological observations. But the truth is that for all experimental data there's a huge range of possible underlying explanations and physicists are stymied.

Is it possible that all of the other mass is in black holes or something another hiding place? The stock answer is NO because black holes are much localized sources of mass and dark matter is very diffuse. I understand that even the massive black holes at the centers of galaxies only make up a small fraction of the mass, often less than a percent.

But black holes are not ruled out. There is still a pretty big region in mass that is not excluded, between where Hawking radiation would be expected to have caused the black holes to have decayed, and masses ruled out by microlensing.

If DM exists, what would a DM "atom" look like?

Since DM is just a theoretical phenomenon arising from a macroscopic level, most speculation as to a DM particle is purely conjecture. Furthermore, theories are for a constituent particle of dark matter has not been detected and cannot be so detected mostly because we do not know what truly understand what constitutes light matter and all of our detectors are built for them. Sure we have measurements, theories of bonds, and plenty of component subatomic components of matter, but scientists have yet to understand the true structure and nature of good ol' regular in your face, matter. Yet here we are conjecturing upon a speculative solution of a cosmological puzzle called DM.

What we know is, DM does not interact via the SF, Strong Force, or EM, Electromagnetism. Whether it reacts with the WF, Weak Force, is still unknown. Scientists can only indirectly measure dark matter distribution through its gravitational influence on ordinary matter and energy, by checking for anomalies in its distribution that cannot be explained by interaction with any of the four known forces of Gravity, EM, SF, and WF.

Up til now, we have only used the spherical harmonic modeshapes with the Real numbers for modeling matter. Let's speculate on the spacetime curvature of an atom that has spherical harmonics which are in the imaginary or complex numbers. As you may recall, the Spherical Harmonics, $Y_{\ell,m}(\theta, \varphi)$, are functions defined on the sphere. They are used to describe the surface of an electron in an atom as like oscillations, like on a soap bubble – vibrating sphere. The spherical harmonics describe non-symmetric solutions to problems with spherical symmetry.

But the $Y_{\ell,m}$'s are complex-valued. The radius of the solutions is the magnitude, and the color shows the phase, of $Y_{\ell,m}(\theta, \varphi)$. These are the numbers on the unit circle: 1 is red, i is purple, -1 is cyan (light blue), and -i is yellow-green. Dark matter may well be the spherical harmonic modeshapes formed where the vibration phase is in the i and –I axis, a toroidal configuration. The toroid has a surface generated by the revolution of any closed plane curve or contour about an axis lying in its plane, hence the effect on the spacetime continuum would be repulsive on all the dark matter atomic surfaces. This could be the cause of complete transparency to any

EM signal that we would normally search for and hence provide the perfect cloaking character.

For each value of ℓ, there are $2\ell + 1$ linearly independent functions $Y_{\ell,m}$, where m = $-\ell, -\ell+1, \ldots, \ell-1, \ell$. I have chosen a different set of $2\ell + 1$ function, as you see below.

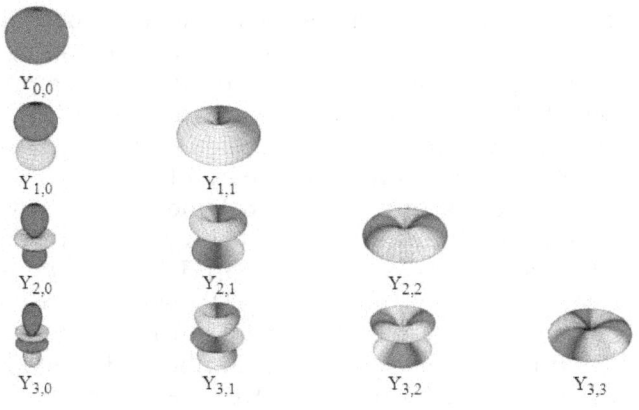

What is interesting about these alternate spherical harmonics from the imaginary component is that they don't come up in the normal vibrating sphere harmonic solutions of the current electron structure, hence matter that we know of, and they are toroidal shaped entities. The toroidal surface encloses a 3-dimensional space as does a sphere. Moreover, the mathematical Integral of the curvature for the toroidal is called the Euler No. and for the torus, this quantity is **zero** for a toroid because the inside curvature is negative and cancels completely with the outside curvature which is positive. A sphere, OTOH, has an Euler No. of **one** because the mathematical Integral of the outside curvature of a sphere is positive.

Since Dark Matter, DM, is not "seeable" or measurable, zero net spacetime curvature, we can speculate further, that DM is another configuration of spacetime with 3 spatial dimensions and a frequency-time but with a twist, canceling out all interaction with real spacetime curvature and hence becoming invisible to our measurements. We proffer that DM "atoms" or particle instead of spherical spacetime are configured in psudospherical shape spacetime or toroidal in shape and hence are everywhere but unseeable and undetectable. Our explanation of this spacetime configuration follows. DM does not emit or interact with EM as we know it or can measure it in any way. DM would have to emit its type of EM, the internal harmonic of its basic structure, to transition to the Minkowski spacetimes, and these transitions will reflect the characteristics of the dark matter as its spacetime makes transitions from DM spacetime to our macro or Minkowski spacetime.

So if I may speculate further, instead of 3-spheres, the DM "atom" is comprised of psuedospherical geometry or imaginary harmonic solutions with boundary entities, without positive curvature boundary surface, which would still warp spacetime, ie add the resistance term in Equation **I-1** above, but its negative curvature outside its boundary would be repulsive or neutral to all other encroaching species of DM, not interacting with any matter particles even slightly as they pass. Since DM atoms would not have to attract curvature, they would not coalesce into larger bodies of DM and hence they would be more evenly distributed than matter in the universe, but still have the overall effect of spacetime expansion slowing that a hidden undetectable dispersed dark mass spacetime curvature in spacetime expansion.

Since most atomic "mass" is inside the atom nucleus, then the DM's bulk mass would likewise reside inside its atomic boundary. The nuclear DM would then also be in likewise smaller psuedospherical configurations. The pseudo-sphere atomic configuration shown would give DM about twice as much "mass" which is what physicists and cosmologists are speculating. However, DM would have completely different membrane boundary harmonics and therefore completely different "electron" or modeshape and harmonic structure and thus invisible to our sensors. A mathematical model depicting a vibrating pseudo-sphere may yield something along the parallel to DM electron structure to aid the speculation. The figures below depict pseudo-sphere, which has some characteristics of a sphere but inverted. It is evident from inspection that the DM atomic structure is repulsive curvature, promoting a wide distribution of Dark matter in the Universe.

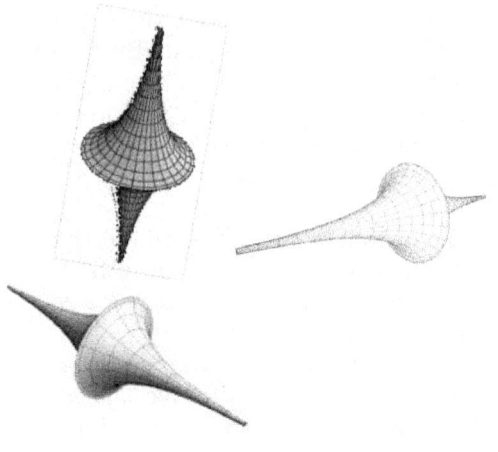

$$\begin{cases} x = a\dfrac{\cos(u+v)}{\operatorname{ch}(u-v)} \\ y = a\dfrac{\sin(u+v)}{\operatorname{ch}(u-v)} \\ z = a(u-v-\operatorname{th}(u+v)) \end{cases} \qquad \begin{cases} x = a\cos u \cos v \\ y = a\cos u \sin v \\ z = a(Gd^{-1}(u) - \sin u) \end{cases}$$

These would comprise the atoms, the atomic surface boundaries of dark matter.

    Why choose a pseudosphere for the basic Dark Matter atom? A pseudosphere surface of radius R is any surface of curvature $-1/R^2$ the negative, concave curvature, of the sphere of radius R, which is a surface of positive curvature $1/R^2$. So if positive curvature is attractive or interactive with other spacetimes, negative curvature will be repulsive and non-interactive with other spacetime entities, thus providing almost total invisibility to us from our current measuring devices and methods.

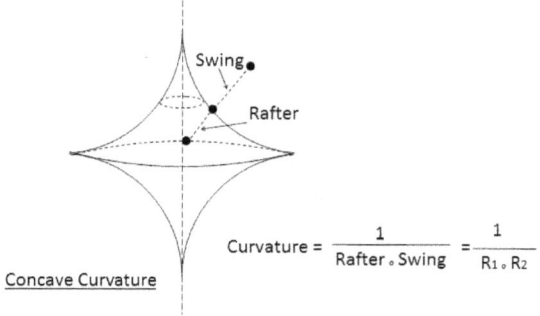

Concave Curvature

Moreover, the repulsive character of the Chronature DM model explains the here-to-fore unexplained repulsive force responsible for the Universe's expansion at an accelerating rate.

From a Universe formation standpoint, the creation of any positive curvature entities, matter as we know it, would likely be in conjunction with the creation of the opposite type of negative curvature spacetime entities, DM as we hypothesize its existence. How does this happen? A quick examination of the surface of the pseudosphere shows that two real spheres, atoms in opposite directions, can be spun off directly opposite to form one puedosphere in the revolution on its axis.

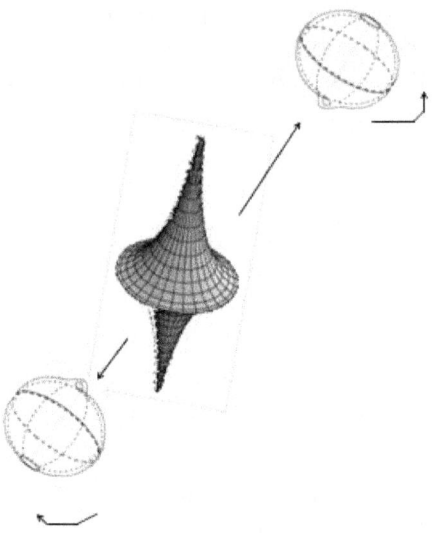

Hence when matter is created, DM is created at the same time. For every atom of DM, we would create an equal and opposite two atoms of regular as-we-know-it, AWKI, matter. Furthermore, the net sum of spacetime surface curvature would be zero and this would comply with the Law of Conservation of Spacetime Curvature, from which all other laws would derive. Also, the pseudosphere has a surface with the appropriate curvature to model a portion of possible hyperbolic space. It would seem unlikely that god would play favorites with Euclidean and Elliptic geometries, forsaking the hyperbolic geometry. Hence the atom, a major component of matter, would be of the non-Euclidian Elliptical geometry with a mathematical opposite non-Euclidian Hyperbolic geometry for the atomic building block of Dark Matter. This would argue that good chance regions and components of the universe are governed by hyperbolic geometry, including analogous counterparts to the internal components of matter, which are comprised of Elliptic geometry. Thus the Universe has a net-zero spacetime curvature, nothing is formed without an equal and opposite entity. This higher and mathematical rule would have a natural subsequent corollary we call Newton's 3rd Law of Equal and Opposite Actions.

The pseudosphere is a three-dimensional surface of constant positive curvature enveloping an origin or center. Opposite but analogous to a sphere with negative Gauss curvature at every point, a positively curved geometry from its center. In counter distinction to the sphere, a pseudosphere has at every point the positively curved geometry of a saddle. Thus it would tend to gravimetrically repel with a tendency to disperse to result in a very diffuse collection of "mass", hence invisible to us through our typical measurements of EM reflection/emission.

Another curious relationship between the two types of spacetime entities of 3-spheres and 4D pseudosphere shapes are volume and surface areas. For a given edge radius R, the area of a pseudosphere is $4\pi R^2$ just as it is for the sphere, and equal in outside boundary curvature. That means that pseudosphere and spheres of the same radius have the same boundary curvature, hence gravitational magnitude. However, the volume of the same radius of a pseudosphere is $2/3\ \pi R^3$ and only half that of a sphere of the same radius, $4/3\ \pi R^3$. Thus psuedosphere "atoms" would have twice the density of spacetime as 3-sphere atoms comprising matter.

Since mass is currently defined above for these types of entities are a function of spacetime curvature and hence volume, it seems appropriate that the surface and curvature surrounding that surface of a spacetime entity be related in mass, twice the volume for the same opposite spacetime curvature. And here we have their masses relationship to be in the proper order of Dark Matter {DM} to matter at two-to-one. Furthermore, the opposite curvature on the pseudosphere would not have the same interaction with EM that matter does, and we would not notice these interactions because all our measuring tools are tuned to EM interaction with matter. The Mobius transformation for a complex pseudosphere would tell us what to look for in terms of interacting entities or structures.

IF DM were in another kind of Spacetime, what could be the closest solution to the EFE as to the metric and structure of such a spacetime?

As it turns out Columbia physicists in pursuit of Dark Matter with a Xenon detector experiments failed time and time again to detect the elusive particle. But the experiments were a fantastic success from an unexpected perspective. So what the Xe Detector DM experiments found was even perhaps more important, even "profoundly influential". The experiments ruled out the existence of many types of dark matter that theorists had proposed. The killer idea that surface was "maybe Einstein miscalculated how gravity works at large scales".

Since Einstein knew of only one kind of spacetime, the Chronature theory of DM from a composite spacetime perspective makes sense.

Above shown and discussed is the Chronature DM particle, but we now have a missing piece that Einstein also missed, Gödel's "revolving universe" solution to the Einstein Field Equations. Looking at the EFE Stress-Energy-Momentum Tensor, from the standpoint of how the DM can be distributed throughout the universe to comply with the EFE we find interesting compatibility between Gödel's solution and what we know and suspect about DM. Since the DM particles are not detectable but still affect the gravity that we see, DM works in Gödel's spacetime in a complementary manner, the symmetry involved Chronature in a composite spacetime. Gödel's spacetime is the missing piece of the Chronature model. It's my DM particles acting in Gödel's revolving universe solution to the EFE. We can provide a roadmap regarding DM in Gödel's "revolving dust" spacetime and so that Chronature's inherent foundation on 4D spacetime is much more complete with Gödel's revolving universe spacetime, perhaps even in the past or at inception.

What is needed is a much more mathematical treatment of the Gödel spacetime universe wrt the little that we know about gravitation generally throughout the Universe.

As stated much above, the Gödel metric is an exact solution to the EFE in which the stress-energy-momentum tensor contains two terms, the first representing the matter density of a homogeneous distribution of swirling dust particles, and the second associated with a nonzero cosmological constant, something which we have ignored in the atomic spacetime thus far. The "dust" particles are representative of DM which are infinitesimally tiny.

Since the energy and momentum of the moving dust particles carry energy and momentum, the moving dust particles will give rise to a gravitational field. Add to that the DM is infinitely tiny "dust particles" that are atoms comprised of non-interacting spacetime curvature entity and composition; conservation equations for stress-energy-momentum tensor will lead to the continuity of a fluid. That is DM acts in fluidic form but without any heat transfer or viscosity characteristics, and entirely from a geometrical standpoint.

The cosmological term definition is somewhat artificial in that the value of the cosmological constant must be carefully chosen to match the density of the dust particles. Since in the Gödel spacetime of DM and perhaps even Dark Energy, there is the fudging parameter of the cosmological constant which becomes important in finding the precise solution or metric.

In all of this is something called Null geodesics. If we examine the past light cone of an event on the axis of symmetry, we find the following picture[47]

---

[47] https://en.wikipedia.org/wiki/G%C3%B6del_metric

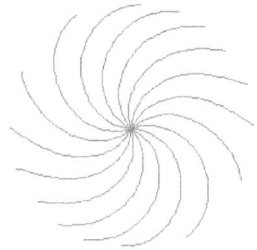 

In the Gödel metric, the null geodesics spiral counterclockwise toward an observer on the axis of symmetry, as shown in the left figure. This looks strikingly similar to the structure of spiral galaxies, shown to the right. This cannot be coincidental. The DM and or Dark Energy must gravitationally affect the nebula structure at formation time and with the null geodesics effect from DM "dust" gravity affects.

The fact that the null geodesics spiral inwards as shown above means that an observer looking radially outwards "sees" nearby DM dust particles, not at their current locations, but their earlier locations. So in compliance with the principle of conservation of spacetime curvature, Gödel metric null geodesics are geometrically straight but in the figures, they appear to spiral only because the coordinates are "rotating" to permit the dust particles to appear stationary. This is just what is expected if the dust particles are rotating about one another.

Concluding Remarks

I'm convinced that Einstein, from deep inside his psyche, knew that his theory of GR and the mysterious set of 4 spacetime dimensions taken as one geometrical entity, one time, and three spatial dimensions, could explain the remaining three known forces. But because of his minimum knowledge of non-Euclidian geometries, Mobius transformations, mathematical inversion, and a host of other mathematical and physical phenomena and lack of an Internet, he was not able to progress his GR theory with Gödel's time loop CTC solution to accurately model atomic-scale spacetime in an atomic model. Instead, he stopped at the boundary of massive objects that caused spacetime to warp causing gravitational forces. Progressing the GR principle of warped spacetime, Chrono Spatial Dynamics, is only an extension of GR into what we call matter, existing in composite spacetimes. Hence, Chroniture is GR applied to other organized entities of spacetime curvatures. Unification is the result of spacetime curvature as it occurs in the physical universe and its dynamics are with the different spacetimes comprised of and defined by different geometries.

Arguably and in line with Occam's Razor, Chronature is the simplest model that fits reality, measurement data and is, therefore, the most plausible of TOE theories. The known four physical forces are all results and property of spacetime curvatures and spacetime curvatures different spacetime curvature sets, comprising all of the particle entities that we have identified in the Standard Model taxonomy. We are sure to yet discover and identify spacetime set entities from such things as dark matter and I proffer that they will take their basic structure from a different configuration of space and time, albeit not necessarily the time or the geometry that we know.

With all of the many theories in Physics propounded today, what distinguishes the theory of Chronature?

Many of modern physics ideas and models are improvable speculation without reasonable plausibility or untestable. This only becomes a problem when the wildly speculative nature of modern physics theories departs from the scientific method. But as the creators of a theory, proponents appear convinced of their ideas on the grounds that they are beautiful or logically or mathematically compelling, despite the impossibility of testing them, much like with String Theory or TOE. They simply sit beyond the practical limits of our ability and capability to test them. They have wandered far beyond the tiniest observable distances or highest possible measurable energies. In the spirit of Woit's "Not Even Wrong" we as physicists can do better, we are wasting valuable intellectual resources focused on those losers.

So what can be done? Some physicists judge the soundness of a theory by using the philosopher Karl Popper's rule of thumb. In the 1930s, Popper drew a line between science and non-science in comparing the work of Albert Einstein with that of Sigmund Freud. Einstein's theory of GR made risky predictions that were eventually proved or agreed with measured phenomena, empirical evidence which substantiated the theory as true. But with Freudian psychoanalysis, any fault of your mother's could be worked into your diagnosis. The theory wasn't "falsifiable", and so as per Popper's rule, it wasn't science.

But at creation, ostensibly the very large bang, theories are speculations, and how does one gain confidence in speculation, a new idea, a new theory? Popper called speculation that did not yield testable predictions "metaphysics," but he considered such activity worthwhile, if it may become testable in the future. Thus if a theory was not "falsifiable" today, it may become so in the future. Speculation should continue but only if the motivation for speculating is clear with the admission that explanations had are not the only possible explanations. It is important to narrow the more viable possibilities to just a few because that would be some progress and worthwhile.

This brings us to Chronature. The motivation to further this theory is its potential to allow us to create experiments and prove them out in a computer, without first having to construct expensive and limiting experimental apparatus to test all theories that are even remotely plausible, but unverifiable from all empirical evidentiary grounds.

Chronature is in complete accordance with Einstein's **correspondence principle** as it produces analogous results from the much older Quantum Field Theory model, analogous as the resultant equations are the same but the physical representation differ, QFT as purely probabilistic and Chronature purely physical. As briefly shown above, Chronature provides that missing link between QT and GR. That missing link is the "Elephant in the room", the physical basis for matter and fields. Chronature offers potentially substantial computational advantages for calculating precise physical characteristics not currently possible, not to mention a leap in understanding the actual real physical structure of matter generally and atomic and molecular bonding specifically.

      Moreover, there is the motivation for a physical model, one solely based on Geometry. The theory of Chronature can be proven out simply by computational models of the structures for the atom and EM as described above, treating particles and subatomic entities as creatures of curved spacetime and calculating properties from those and their interactions. The other possibilities and explanations have so far proven incomplete or too tough to prove experimentally or both.

Appendix

So what? It's yet another theory

A realist would say that theories are of little value unless they can be put to practical use or give us an understanding of nature and hope for some benefit to mankind. Therefore the first goal and test of the Chronature model's usefulness are to show that it can be applied. Chronature models should be directly applicable in cracking water economically. This single breakthrough could solve the world's top five problems with solutions. 1) reversal of Climate Change, 2) virtually inexhaustible clean renewable energy source, 3) cessation of oil geopolitics, and 4) elimination of air pollution and water contamination from fossil fuel production and use, 5) heavy metal contamination from industrial processes and products.

The means to crack water is predicated by the calculation of the physical parameters of specific atomic characteristics and attributes that are important in bonding Hydrogen and Oxygen from their characteristic spacetime curvatures, resonances, and the many physical curvature dynamics. The atomic parameters of specific atom modeshapes and curvatures, spectrum for these precise modes, interaction distances, modes of interaction, and motions of rotation-vibration-translation-deformation, from a spacetime curvature basis. What is needed are the physical vibrating sphere atom model curvature calculations for atomic bonds in the water molecule and its constituents in question. We can then proceed to use models to engineer H-O bond separation using nano-structures, direct induction, resonant magnetic induction, electromagnetic radiation in the form of microwaves, transport through nanostructures, Femto pulse lasers, and electrical conduction through natural media, such as aqueous solutions with salts through electrolysis, graphene nets, etc, but tailored to the precise bond spacetime curvature of H-O interaction. Avoidance of the current brut force non-spontaneous disassociating reaction is a minimum for the goal of producing a cascading "falling" away of a Hydrogen atom from the Oxygen atom bond to be conducted to separate volumes.

For the reasons above, applying the Chronature model to understand chemical bonds for "unbonding" should be our first goal. We must first understand charge, voltage, current, and conduction from a Chronature spacetime curvature perspective applied to oxygen and hydrogen to design a system to break the bonds using organized spacetime curvature resonance and finesse instead of the current brute force methods which are energy expensive because they are subject to the randomness and statistics of the 1st, 2nd and 3rd laws of Thermodynamics.

Why use the Chronature model?

We have the Einstein Field Equations, EFE from GR, and QM models to make all manner of calculations. The fact that solutions to the vibrating sphere equations have been known even before GR, and that with some careful choices spacetime curvatures of 4D surfaces are easily computable means that we can calculate the curvature of spacetime without the EFE. These along with the QM method's complexities EFE makes current computational methods for solving real problems untenable. Hence Chronature model can be incorporated using simple models to generate the numbers for bond forces from spacetime curvature geometry between atoms and molecules, as well as physical plots to aid in the understanding of reality. Having useable computer models for simple to complex atomic and molecular structures cannot be overstated, in obtaining calculations that can aid in cracking the H-O bond cheaply.

    In addition, we now have another major advantage that is we now have an actual physical model, which we can apply to predict what we measure in a computer model without translation from another less accurate statistical model. With computer simulations, we will calculate the size, time, resonance of atomic and molecular structures with actual physical bonding parameters. Using principles of resonance we should be able to produce physical molecular cascading resonance. As a domino can knock over an adjacent domino 1.5x larger than itself, a series of 13 dominoes each 1.5x the previous can amplify an initial push 2 billion times its initial push. With precise engineering numbers of the bond parameters via spacetime curvature, our objective will be to build an apparatus that will provide a little push and will cascade/amplify that little perturbation to the magnitude necessary to break H-O bonds without the brute force currently used. And that should be with less energy than what can be derived from the recombination under combustion conditions. As my friend Jay exhorts, "spacetime is of the essence."

Homework Assignment

Calculating Spacetime Curvature for Atomic Surfaces

Assume that the atom is another type of spacetime, bounded by a vibrating sphere partitioning two disparate spacetimes. The vibrating sphere boundary surface has a closed-form solution, calculable spacetime curvature at every point, and calculable without the use of Einstein Field Equations. The Einstein Field Equations are formidable and overkill since they were made for every aspect of every

$$G_{\mu\nu} + \Lambda g_{\mu\nu} = \frac{8\pi G}{c^4} T_{\mu\nu}$$

geometry.

Here we propose an alternate means to calculating the spacetime curvature for matter at points of interest, in this case, the atomic bonds between Hydrogen and Oxygen. Similar to deriving the acceleration of gravity by using the spacetime geometry of a parabola/paraboloid above, or hyperboloid or ellipse, we can in a similar fashion derive all the forces and acceleration fields between atoms and among molecules, by using the spacetime geometry of those entities to calculate the spacetime curvatures at maximum curvature points and derive the accelerations from that geometrical data alone. This begins with finding the osculating circle radius for the atom harmonic modeshapes, "electrons", and calculating the acceleration fields from the inverse of the osculating radius of these shapes, which are known as the actual approximate dimensions.

Since the atomic boundary is a spacetime surface, we would calculate the curvature for use in atomic bond forces of attraction for values at specific points in space and time. That is, we are looking for the curvature calculated from the smallest osculating circle radius at the atomic harmonic modeshape lobe tips, much like that of the paraboloid of acceleration of gravity field on the Earth's surface in spacetime above. The difference is that at the atomic level, the spacetime curvature components due to spatial curvature are magnitudes greater than curvature, acceleration force, from temporal components.

Using what has been developed by the Quantum Mechanics camp, the solutions for the vibrating sphere as our spherical atom boundary are harmonics which at any time, sum to the total spacetime curvature at any point in that time. However, the spacetime curvature takes on the specific 3D surface curvatures of the mode shape at resonances in 3D space. Using Separation of Variables, solutions to the Vibrating Sphere Equation without Time variable, many available on the web, are found below. These solutions given below are serendipitously identical to the atomic physics model and theory derived from the Bohr atom QMT modifications, and used here:

$\phi \rightarrow e^{im\phi}$

$\theta \rightarrow f_{lm}(\theta) \rightarrow$ Legendre Functions

$f_{lm}(\theta) g(\phi) \rightarrow Y_{lm}(\theta, \phi) \rightarrow$ Spherical Harmonics

$R_{nl}(r) \rightarrow$ Laguerre polynomials

$\psi_{nlm}(r, \theta, \phi) = C_{nlm} R_{nl}(r) f_{lm}(\theta) g_m(\phi)$

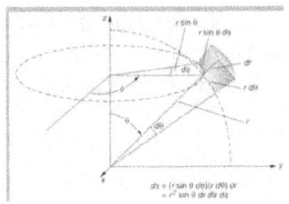

Spherical and radial functions for specific harmonics for l, m, and n, l quantum numbers respectively:

$Y_{00} = \dfrac{1}{\sqrt{4\pi}}$ $R_{10} = 2\left(\dfrac{Z}{a_0}\right)^{\frac{3}{2}} e^{-Zr/a_0}$

$Y_{11} = -\sqrt{\dfrac{3}{8\pi}} e^{i\phi} \sin\theta$ $R_{21} = \dfrac{1}{\sqrt{3}} \left(\dfrac{Z}{2a_0}\right)^{\frac{3}{2}} \left(\dfrac{Zr}{a_0}\right) e^{-Zr/2a_0}$

$Y_{10} = \sqrt{\dfrac{3}{4\pi}} \cos\theta$ $R_{20} = 2\left(\dfrac{Z}{2a_0}\right)^{\frac{3}{2}} \left(1 - \dfrac{Zr}{2a_0}\right) e^{-Zr/2a_0}$

$Y_{22} = \sqrt{\dfrac{15}{32\pi}} e^{2i\phi} \sin^2\theta$ $R_{32} = \dfrac{2\sqrt{2}}{27\sqrt{5}} \left(\dfrac{Z}{3a_0}\right)^{\frac{3}{2}} \left(\dfrac{Zr}{a_0}\right)^2 e^{-Zr/3a_0}$

$Y_{21} = -\sqrt{\dfrac{15}{8\pi}} e^{i\phi} \sin\theta \cos\theta$ $R_{31} = \dfrac{4\sqrt{2}}{3} \left(\dfrac{Z}{3a_0}\right)^{\frac{3}{2}} \left(\dfrac{Zr}{a_0}\right) \left(1 - \dfrac{Zr}{6a_0}\right) e^{-Zr/3a_0}$

$Y_{20} = \sqrt{\dfrac{5}{16\pi}} (3\cos^2\theta - 1)$ $R_{30} = 2\left(\dfrac{Z}{3a_0}\right)^{\frac{3}{2}} \left(1 - \dfrac{2Zr}{3a_0} + \dfrac{2(Zr)^2}{27a_0^2}\right) e^{-Zr/3a_0}$

(a) Using QM's solutions of the wavefunction above for a sphere harmonic modeshape representing an H atom electron "orbital", the same as a vibrating sphere model, with radius a_0, solve for the four lowest non-ground state spherical harmonics (n, l, m_l) = {(2,0, 0), (2, 1, 0), (m_l =1, ± 1) for a vibrating sphere of radius a_0.

(b) Converting the two complex spherical harmonics for l = 0, m_l = 0 ($1S^1$ shell), and for l = 1, m_l = ± 1 into real normalized functions that represent 1s, $2p_x$, $2p_y$, $2p_z$. Combine these with the appropriate "radial wavefunctions" harmonic modeshape (Principle Quantum Number n=2 and Bohr radius a_0 - H atom)

444

to derive the radial component solutions for each harmonic modeshape for the H atom.

where the Bohr radius is

$$a = \frac{4\pi\epsilon_0 \hbar^2}{me^2},$$

(c) Obtain the total modeshape solution for the s and p harmonics using.

$$\Psi_{n\ell m}(r,\theta,\phi) = \sqrt{\left(\frac{2}{na_0}\right)^3 \frac{(n-\ell-1)!}{2n[(n+\ell)!]}} e^{-r/na_0} \left(\frac{2r}{na_0}\right)^\ell L_{n-\ell-1}^{2\ell+1}\left(\frac{2r}{na_0}\right) \cdot Y_\ell^m(\theta,\phi)$$

where $a_0 = 4\pi\epsilon_0\hbar^2/m_e e^2$ is the Bohr radius, $L_{n-\ell-1}^{2\ell+1}$ are the generalized Laguerre polynomials of degree $n - \ell - 1$, $n = 1, 2, \ldots$ is the principal quantum number, $\ell = 0, 1, \ldots n - 1$ the azimuthal quantum number, $m = -\ell, -\ell+1, \ldots, \ell-1, \ell$ the magnetic quantum number. Hydrogen-like atoms have very similar solutions.

(d) Differentiate the s and 2p radial wavefunctions of a hydrogen atom and locate the two extrema in atom radius, or this can be done by inspection, ie select the modeshape leaf pod endpoints, ie. where the radius of curvature is smallest (highest curvature) for any point on the leaf. Then, write an expression for the radial distribution function and determine for the H atom (i) the p harmonic modeshape curvature at the largest radius of the s and p harmonic modeshapes ie find the osculating circle radius at the largest mode shape radius, and use $1/(R_{osc})^2$ to calculate the curvature at that extrema point. See the diagram below for points where the curvature on the s and p mode shapes will be calculated. Because they are inverse radius squared, the maximum curvature would occur in the smallest osculating circles, tangent to the node extremes of leaf tips. The principle max curvature point tangents would be normal to each other as well as normal to the modeshape tip surface. The modeshape pods or leaves are symmetrical about their outward axis, and this simplifies obtaining the osculating principle axis radii to one radius, as we found in the calculation of the earth's gravity from the spacetime paraboloid curve from a few measured points. In our case here, the geometrical curves are closed-form solutions and well know for the vibrating sphere of the variable radius by elements.

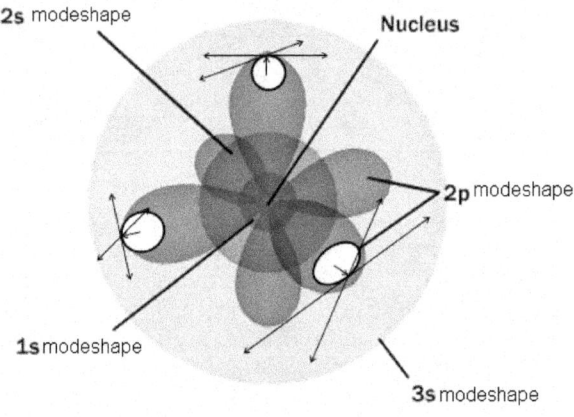

(f) What is the maximum curvature, ie. at point smallest tangent osculating sphere radius, for the Hydrogen atom 1s and $2p_n$ spherical harmonic modeshapes?

(g) Calculate the max curvatures for the $1s^2 2s^2\ 2p^4$ modeshape surfaces as well. hint: find the osculating circle radius at the lobe tip.

(h) The Time component has largely been removed from the vibrating sphere solutions as they were not useful for QM. Derive and use the vibrating sphere solutions for the Time dimension, find the frequency of the harmonics at $1s^1$ (H) and $1s^2 2s^2\ 2p^4$ (O)

Lastly, in spherical coordinates the metric or line element takes the form:

$$ds^2 = c^2 dt^2 - dr^2 - r^2 d\theta^2 - r^2 \sin^2(\theta) d\phi^2$$

This is the coordinate system which is useful to apply the position curves or modeshapes to calculate field strength from spacetime curvature alone.

Answers/clues for a -f

(a) $Y_{lm}(\theta, \phi)$, for azimuthally and magnetic harmonic index l = 0, l = 1 and m = 0, ±1

$Y_{lm}(\theta, \phi) = Y_l^{\pm m}(\theta, \phi)$, first convert two complex spherical harmonics into their real normalized spherical harmonic solution for "azimuthal" and magnetic" quantum numbers l and m.

$Y_0^0(\theta, \phi) = 1/(4\pi)^{1/2}$ 1S -
First excited harmonic l=0, m=0. Note modeshape only depends on the radius as this is an S shell, the max radius of curvature from tangent osculating sphere should be that the solution radius for 1S harmonic.

$Y_1^0(\theta, \phi) = \underline{(3/4\pi)^{1/2} \cos(\theta)}$ \hfill 2P$_Z$ -
note curvature is max at $\theta = 0$ and $n\pi$

$Y_1^{+1}(\theta, \phi) = (3/8\pi)^{1/2} (-\sin(\theta)\, e^{i\phi})$ \hfill 2P$_X$ -
The real part is modeshape $\pm$X axis

$Y_1^{+1}(\theta, \phi) = (3/8\pi)^{1/2} (-\sin(\theta)(\cos\phi \pm i\sin\phi))$

$\quad$ Re($Y^{+1}_1(\theta, \phi)$) = $\underline{(-/+)(3/8\pi)^{1/2} (\sin\theta \cos\phi)}$

$Y_1^{-1}(\theta, \phi) = (3/8\pi)^{1/2} \sin(\theta)\, e^{-i\phi}$ \hfill 2P$_Y$ -
The Imaginary part is the $\pm$ Y axis
$Y_1^{-1}(\theta, \phi) = (3/8\pi)^{1/2} (\sin(\theta))(\cos\phi)(-/+ i\sin\phi)$

$\quad$ Im($Y_1^{-1}(\theta, \phi)$) = $\underline{(-/+)(3/8\pi)^{1/2} (\sin(\theta)(\pm \sin\phi)}$

(b) Now solving for the S and P harmonic radial components, Z is generally the number of proton excitation of the atom boundary from inside of the atom, but for H higher harmonic modeshapes, higher "energies", the Z's increase in the order of excitation from photons or other particles from outside the H atom boundary.

$$R_n^1(r) = 2/(3)^{1/2}(Z/2a_0)^{3/2} (Zr/2a_0)\, e^{-(Zr/2a0)}$$

$$R_2^0(r) = 2/(3)^{1/2}(2/2a_0)^{3/2} (2r/2a_0)\, e^{-(2r/2a0)}$$

$$= \underline{2/(3)^{1/2} (1/a_0)^{3/2} (r/a_0)\, e^{-(r/a0)}}$$

$$R_2^{\pm 1}(r) = (2/(3)^{1/2}(2/2a_0)^{3/2} (Zr/2a_0)\, e^{\pm (2r/2a0)}$$

$$R_2^{\pm 1}(r) = 2/(3)^{1/2}(1/a_0)^{3/2}\,(r/a_0)\,e^{\pm(r/a0)}$$

(c) What is the maximum curvature for the following spherical harmonic modeshapes?

In order to find the acceleration field strength we must find the osculating circle radius at the modeshape lobe tips. This will represent the maximum acceleration due to the change in spacetime geometry for each spatial axis. Assume that the principle, azimuthal, and magnetic quantum numbers are n, l, and m. Also remember that the radius is for H and that the typical QM model assigns the 1S shell to Hydrogen.

$Y_{nlm}(\theta, \phi) = 2/(3)^{1/2}(Z/2a_0)^{3/2}(Zr/2a_0)\,e^{-(Zr/2a0)}\,Y_l^{\pm m}(\theta, \phi)$,

For the 1S harmonic
$\quad Y_{nlm}(\theta, \phi) = Y_{200}(\theta, \phi) = 2/(3)^{1/2}(2/2a_0)^{3/2}(2r/2a_0)\,e^{\pm(2r/2a0)}\,1/(4\pi)^{1/2}$
$\quad = 1/(3\pi)^{1/2}(1/a_0)^{3/2}(r/a_0)\,e^{\pm(r/a0)}$

For the P shell harmonics

$\quad Y_{nlm}(\theta, \phi) = Y_{210}(\theta, \phi)$
$= 2/(3)^{1/2}(1/a_0)^{3/2}(r/a_0)\,e^{\pm(r/a0)} \times (3/4\pi)^{1/2}\cos(\theta)$
$= (1/\pi)^{1/2}(1/a_0)^{3/2}(r/a_0)\,e^{\pm(r/a0)} \times \cos(\theta)$

$\quad Y_{nlm}(\theta, \phi) = Y_{211}(\theta, \phi)$
$= 2/(3)^{1/2}(2/2a_0)^{3/2}(2r/2a_0)\,e^{-(2r/2a0)} \times (-/+)(3/8\pi)^{1/2}(\sin\theta\,\cos\phi))$

$Y_{nlm}(\theta, \phi) = Y_{21-1}(\theta, \phi) = 2/(3)^{1/2} (2/2a_0)^{3/2}$
$(2r/2a_0) e^{-(2r/2a_0)} \times (\pm)(3/8\pi)^{1/2} (\sin(\theta)(\sin\phi)(\pm)(3/2\pi)^{1/2}$
$(1/a_0)^{3/2} (r/a_0) e^{-(r/a_0)} \times (\sin(\theta)(\sin\phi)$

I found these tables useful to visualize the harmonic indexes and their effect on modeshape, the spacetime geometry curves for particular atomic harmonics. The important quantities to be extracted from these solutions are the maximum curvatures, generally, those points sharpest or with the smallest osculating radii and generally found at the leaf ends

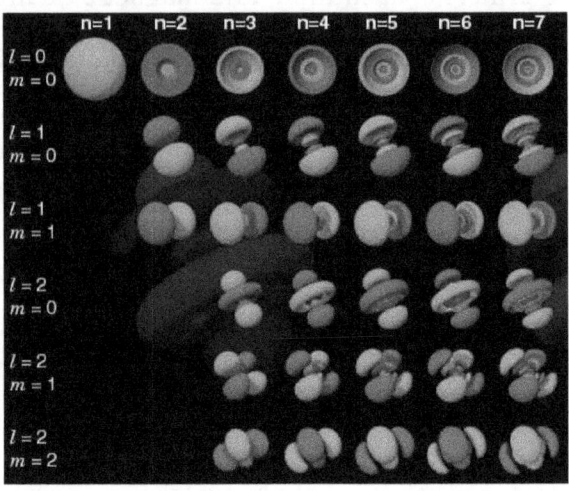

l:	$P_\ell^m(\cos\theta)\cos(m\varphi)$						$P_\ell^{|m|}(\cos\theta)\sin(|m|\varphi)$					
0 s												
1 p												
2 d												
3 f												
4												
5												
6												
m:	6	5	4	3	2	1	0	-1	-2	-3	-4	-5	-6

(h) Using the standard wave equation, convert to spherical coordinates, solve the PDE for the time solutions for natural frequencies, solve for atomic natural harmonic frequencies at the modeshapes derived above for H and O

The basic wave equation:

$$\frac{\partial^2 f(x,t)}{\partial x^2} = \frac{1}{c^2}\frac{\partial^2 f(x,t)}{\partial t^2}$$

The basic wave equation with a resistance, inertia, term:

$$-\frac{1}{c^2}\frac{\partial^2}{\partial t^2}\psi = -\nabla^2\psi + \frac{m^2 c^2}{\hbar^2}\psi$$

Bibliography

1. Lee Smolin, *"The Trouble With Physics"*, pg 4, pg 41

2. QED, Richard P. Feynman, pg 98, *"Every particle in Nature has an amplitude to move backwards in time, and therefore has an anti-particle."*

3. Tristan Needham, *Visual Complex Analysis*, pg 127

4. Lisa Randall, *Warped Passages*, pg 300, pg 352

5. Lewis Carroll Epstein, *Relativity Visualized*, pg 150

6. Wikipedia, http://en.wikipedia.org/wiki/Schr%C3%B6dinger_equation

7. Donald H. Menzel, *Mathematical Physics*, pg 213

8. Peter Woit, *Not Even Wrong*,

9. Richard Feynman, *"QED – A Strange Theory of Light and Matter"*

10. Richard P. Feynman, *Six Easy Pieces*, pg 44

11. D.M.Y. Sommerville, *The Elements of Non-Euclidean Geometry*, pg 92

13. Wikipedia *"Laplace's Spherical Harmonic"*

14. H.M.Schey, *Div, Grad, Curl, and all That*, pg 5

15. W.K. Clifford, *The Common Sense of the Exact Sciences*

16. http://en.wikipedia.org/wiki/Atomic_orbital

17. Hans A Bethe and Philip Morrison, *Elementary Nuclear Theory*, pg 162-173

18. Stefan Kulczycki, **Non-Euclidean Geometry**, pg 170-172

19. http://www.universetoday.com/83759/astronomy-without-a-telescope-unreasonable-effectiveness/

20. Atmospheric Optics, http://www.atoptics.co.uk/droplets/corform.htm

21. Solitary Waves" from Scientific American, Volume 80, pg 358

22. Michio Kaku, **Hyperspace**, pg 146-147

23, Max Tegmark, *Our Mathematical Universe*, pg 165

24, Space and Time - Minkowski's Papers on Relativity, Minkowski Institute Press, pg 24

25. Keith Devlin, "The Millennium Problems", pg 179-182

About the Author

    Walt Froloff is a recovering engineer, physicist, inventor, and practicing attorney residing in Northern California. He holds patents in the fields of Artificial Intelligence, Business Intelligence, Mobile Advertising, Intelligent Engines, Air Hybrid Engine, Air Impulse Engine, Medical Devices and Consumer Products.

    He has written *Irrational Intelligence,* and *Learning Elephant.* His mantra is, "If you think it's impossible, you're thinking to much – go by feel."

    More about how to harness the power of feelings intelligence in electronic devices can be found at FeelingsIntel.com. His smart engine technology can be referenced on InspirEngine.com. His legal practice can be referenced on PatentAlchemy.com. He is involved in introducing mobile ad business models at Feelmobi.com applying Feelings Intelligence on mobile apps. PerchCam.com and BiopathAI.com and AgSensAI.com are the latest inventions Walt is pursuing.

www.ingramcontent.com/pod-product-compliance
Lightning Source LLC
Chambersburg PA
CBHW061502180526
45171CB00001B/6